COMPUTER GRAPHICS FROM SCRATCH

COMPUTER GRAPHICS FROM SCRATCH

A Programmer's Introduction to 3D Rendering

by Gabriel Gambetta

no starch press

San Francisco

25 24 23 22 21 1 2 3 4 5 6 7 8 9

ISBN-13: 978-1-7185-0076-1 (print)
ISBN-13: 978-1-7185-0077-8 (ebook)

Publisher: William Pollock
Executive Editor: Barbara Yien
Production Manager: Rachel Monaghan
Production Editor: Kassie Andreadis
Developmental Editor: Alex Freed
Cover Illustrator: Rob Gale
Interior Design: Octopod Studios
Technical Reviewer: Alejandro Segovia Azapian
Copyeditor: Gary Smith
Proofreader: Elizabeth Littrell
Indexer: Elise Hess

For information on book distributors or translations, please contact No Starch Press, Inc. directly:
No Starch Press, Inc.
245 8th Street, San Francisco, CA 94103
phone: 1-415-863-9900; info@nostarch.com
www.nostarch.com

Library of Congress Cataloging-in-Publication Data
Names: Gambetta, Gabriel, 1980- author.
Title: Computer graphics from scratch : a programmer's introduction to 3D rendering / Gabriel Gambetta.
Description: San Francisco : No Starch Press, [2021] | Includes index.
Identifiers: LCCN 2020056364 (print) | LCCN 2020056365 (ebook) | ISBN 9781718500761 (print) | ISBN 9781718500778 (ebook)
Subjects: LCSH: Computer graphics.
Classification: LCC T385 .G3524 2021 (print) | LCC T385 (ebook) | DDC 006.6--dc23
LC record available at https://lccn.loc.gov/2020056364
LC ebook record available at https://lccn.loc.gov/2020056365

To my dad (1947–2007), architect and self-taught programmer, who got me started on this path.

My dad, my two-and-a-half-year-old self, and the ZX81.

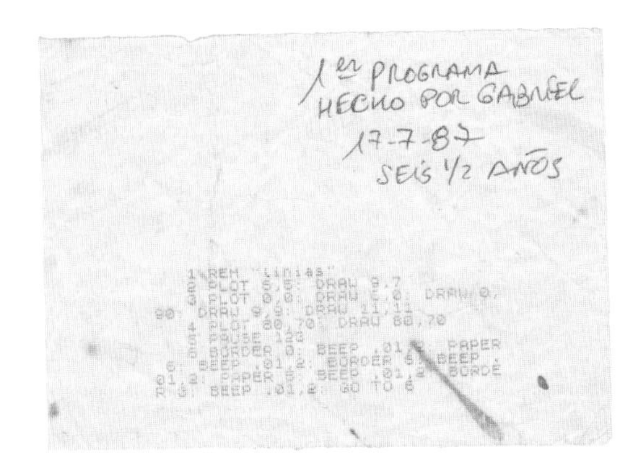

My earliest documented program ever, written at six-and-a-half years old, drew some lines on the screen of my ZX Spectrum+.

About the Author

Gabriel Gambetta started coding games around the age of 5 on a ZX Spectrum. After studying computer science and working at a respectable local company in his native Uruguay, he started a game development company and ran it for 10 years while teaching computer graphics at his alma mater. More recently, Gambetta has been working at Google Zürich since 2011, except for a stint as an early engineer at London-based multiplayer game tech unicorn Improbable, and a year in Madrid focusing on acting and filmmaking.

About the Technical Reviewer

Alejandro Segovia Azapian is a software engineer with 14+ years of experience in computer graphics. He has worked for several industry-leading companies in the field of 3D graphics including Autodesk, Electronic Arts, PDI/DreamWorks, and WB Games, across a variety of realtime graphics projects spanning apps, games, game engines, and frameworks. Alejandro currently works in the GPU Software group at a leading consumer electronics company based in Cupertino, California.

BRIEF CONTENTS

CONTENTS IN DETAIL

ACKNOWLEDGMENTS

Few books happen overnight; the one you're about to read has been almost 20 years in the making. As you might suspect, many people have been part of its story, in one way or another, and I want to thank them. In chronological order:

Omar Paganini and Ernesto Ocampo Edye As the Dean of the School of Engineering and the Director of Computer Science at Universidad Católica del Uruguay, they put considerable trust in me by letting me take the reins of Computer Graphics when I was but a fourth-year student, and by letting me completely reshape its curriculum in the way I thought best. Fellow professor **Roberto Lublinerman** was a great mentor throughout my first year of teaching.

My students from 2003 to 2008 Besides being the unwitting guinea pigs of my continuously evolving teaching methodology, they accepted and respected a professor barely a year older than them (and, in some cases, younger than them). The joy in their faces when they created their first raytraced images made it all worth it.

Alejandro Segovia Azapian A student turned teaching assistant turned friend, his input has helped me evolve the material over time; having been a tiny part of his subsequent, very successful professional career specialized in realtime rendering and performance optimization fills me with pride. He was also a technical reviewer of this book, and his contributions ranged from fixing typos to suggesting deep structural improvements of some chapters.

JC Van Winkel He did his own editing pass and came up with a lot of valuable suggestions for improvement.

The readers of *Hacker News* My lecture notes, diagrams, and demos made the front page of *Hacker News*, and attracted considerable attention—including that of No Starch Press. If this hadn't happened, this book might have never existed.

Bill Pollock, Alex Freed, Kassie Andreadis, and the entire No Starch Press team They guided me through cleaning up and reshaping my lecture notes and diagrams, which in my mind were essentially ready to be published as a book, into an actual book. They took the raw materials to a whole new level; I had no idea this took so much work and effort, and Alex, Kassie, and the team did a stellar job. My name is the only one on the cover, but make no mistake, this was a group effort.

INTRODUCTION

Computer graphics is a fascinating topic. How do you go from a few algorithms and some geometric data to the special effects for movies like *Star Wars* and *The Avengers*, animated movies like *Toy Story* and *Frozen*, or the graphics of popular video games like *Fortnite* or *Call of Duty*?

Computer graphics is also a frighteningly broad topic: from rendering 3D scenes to creating image filters, from digital typography to simulating particle systems, there are a multitude of disciplines that can be thought of as part of computer graphics. One book couldn't hope to cover all these subjects; it would take a library. This book focuses exclusively on the topic of rendering 3D scenes.

Computer Graphics from Scratch is my humble attempt to present this one slice of computer graphics in an accessible way. It is written to be easily understood by high-school students, while staying rigorous enough for professional engineers. It covers the same topics as a full university course—it is, in fact, based on my years of teaching the subject at university.

Who This Book Is For

This book is for anyone with an interest in computer graphics, from high-school students to seasoned professionals.

I have made a conscious choice to favor simplicity and clarity in the presentation. This is reflected in the choice of ideas and algorithms within the book. While the algorithms are industry-standard, whenever there's more than one way to achieve a certain result, I have chosen the one that is easiest to understand. At the same time, I've put considerable effort into making sure there's no hand-waving or trickery. I tried to keep in mind Albert Einstein's advice: "Everything should be made as simple as possible, but no simpler."

There's little prerequisite knowledge and no hardware or software dependencies. The only primitive used in this book is a method that lets us set the color of a pixel—as close to *from scratch* as we can get. The algorithms are conceptually simple, and the math is straightforward—at most, a tiny bit of high-school trigonometry. We also use some linear algebra, but the book includes a short appendix presenting everything we need in a very practical way.

What This Book Covers

This book starts from scratch and builds up to two complete, fully functional renderers: a raytracer and a rasterizer. Although they follow very different approaches, they produce similar results when used to render a simple scene. Figure 1 shows a comparison.

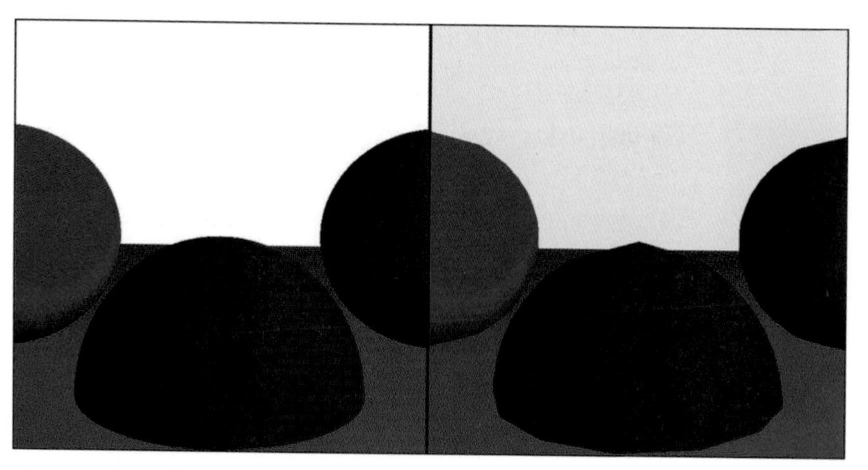

Figure 1: A simple scene rendered by the raytracer (left) and the rasterizer (right) developed in this book.

While the features of the raytracer and rasterizer have considerable overlap, they are not identical, and this book explores their specific strengths, some of which can be seen in Figure 2.

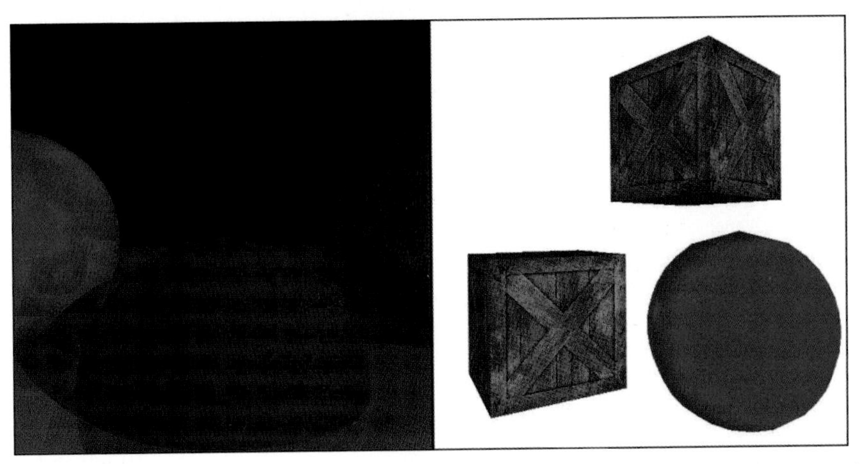

Figure 2: The raytracer and the rasterizer have their own unique features. Left: raytraced shadows and recursive reflections; right: rasterized textures.

The book provides informal pseudocode throughout the text, as well as links to fully working implementations written in JavaScript that can run directly in any web browser.

Why Read This Book?

This book should give you all the knowledge you need to write software renderers. It does not make use of, or teach you how to use, existing rendering APIs such as OpenGL, Vulkan, Metal, or DirectX.

Modern GPUs are powerful and ubiquitous, and few people have good reason to write a pure software renderer. However, the experience of writing one is valuable for the following reasons:

Shaders are software. The first, ancient GPUs of the early 1990s implemented their rendering algorithms directly in hardware, so you could use them but not modify them (which is why most games from the mid-1990s look so similar to each other). Today, you write your own rendering algorithms (called *shaders* in this context) and they run in the specialized chips of a GPU.

Knowledge is power. Understanding the theory behind the different rendering techniques, rather than copying and pasting half-understood fragments of code or cargo-culting popular approaches, lets you write better shaders and rendering pipelines.

Graphics are fun. Few areas of computer science provide the kind of instant gratification offered by computer graphics. The sense of accomplishment you get when your SQL query runs just right is *nothing* compared to what you feel the first time you get raytraced reflections right. I taught computer graphics at university for five years, and I often wondered why I enjoyed teaching the same thing semester after semester for so long; in the end, what made it worth it was seeing the faces of my students light up and seeing them use their first rendered scenes as their desktop backgrounds.

About This Book

This book is divided into two parts, *Raytracing* and *Rasterization*, corresponding to the two renderers we are going to build.

The first chapter introduces some basic knowledge necessary to understand these two parts. I suggest you read the chapters in order, but both parts of the book are self-contained enough that they can be read mostly independently.

Here's a brief overview of what you'll find in each chapter.

Chapter 1: Introductory Concepts We define the *canvas*, the abstract surface we'll be drawing on, and `PutPixel`, our only tool to draw on it. We also learn to represent and manipulate colors.

Part I: Raytracing

Chapter 2: Basic Raytracing We develop a basic raytracing algorithm capable of rendering a few spheres, which look like colored circles.

Chapter 3: Light We establish a model of how light interacts with objects and extend the raytracer to simulate light. The spheres now look like spheres.

Chapter 4: Shadows and Reflections We improve the appearance of the spheres: they cast shadows on each other and can have mirror-like surfaces where we can see reflections of other spheres.

Chapter 5: Extending the Raytracer We present an overview of additional features that can be added to the raytracer, but which are beyond the scope of this book.

Part II: Rasterization

Chapter 6: Lines We start from a blank canvas and develop an algorithm to draw line segments.

Chapter 7: Filled Triangles We reuse some core ideas from the previous chapter to develop an algorithm to draw triangles filled with a single color.

Chapter 8: Shaded Triangles We extend the algorithm from the previous chapter to fill our triangles with a smooth color gradient.

Chapter 9: Perspective Projection We take a break from drawing 2D shapes to look at the geometry and math we need to convert a 3D point into a 2D point we can draw on the canvas.

Chapter 10: Describing and Rendering a Scene We develop a representation for objects in the scene and explore how to use perspective projection to draw them on the canvas.

Chapter 11: Clipping We develop an algorithm to remove the parts of the scene that the camera can't see. Now we can safely render the scene from any camera position.

Chapter 12: Hidden Surface Removal We combine perspective projection and shaded triangles to render solid-looking objects; for this to work correctly, we need to ensure distant objects don't cover closer objects.

Chapter 13: Shading We explore how to apply the lighting equation developed in Chapter 3 to entire triangles.

Chapter 14: Textures We develop an algorithm to "paint" images on our triangles as a way to fake surface detail.

Chapter 15: Extending the Rasterizer We present an overview of features that can be added to the rasterizer, but which are beyond the scope of this book.

Appendix: Linear Algebra We introduce the basic concepts from linear algebra that are used throughout this book: points, vectors, and matrices. We present the operations we can do with them and provide some examples of what we can use them for.

About the Author

I'm a senior software engineer at Google. In the past, I've worked at Improbable (*http://improbable.io*), who have a good shot at building the Matrix for real (or at the very least revolutionizing multiplayer game development), and at Mystery Studio (*http://mysterystudio.com*), a game development company I founded and ran for about a decade and which released almost 20 games you've probably never heard of.

I taught computer graphics for five years at university, where it was a semester-long third-year subject. I am grateful to all of my students, who served as unwitting guinea pigs for the materials that inspired this book.

I have other interests besides computer graphics, engineering-related and otherwise. See my website, *http://gabrielgambetta.com*, for more details and contact information.

1

INTRODUCTORY CONCEPTS

A raytracer and a rasterizer take very different approaches to rendering a 3D scene onto a 2D screen. However, there are a few fundamental concepts that are common to both approaches.

In this chapter, we'll explore the canvas, the abstract surface on which we'll render our images; the coordinate system we'll use to refer to pixels on the canvas; how to represent and manipulate colors; and how to describe a 3D scene so our renderers can work with it.

The Canvas

Throughout this book, we'll be drawing things on a *canvas*: a rectangular array of pixels that can be individually colored. Whether this canvas is shown on a screen or printed on paper is irrelevant to our purposes. Our goal is to represent a 3D scene on a 2D canvas, so we'll focus on rendering to this abstract, rectangular array of pixels.

We'll build everything in this book out of a single, very simple function, which assigns a color to a pixel on the canvas:

```
canvas.PutPixel(x, y, color)
```

This method has three arguments: an x coordinate, a y coordinate, and a color. Let's focus on the coordinates for now.

Coordinate Systems

The canvas has a width and a height, measured in pixels, which we'll call C_w and C_h. We need a coordinate system to refer to its pixels. For most computer screens, the origin is at the top left; *x* increases toward the right of the screen, and *y* increases toward the bottom, as in Figure 1-1.

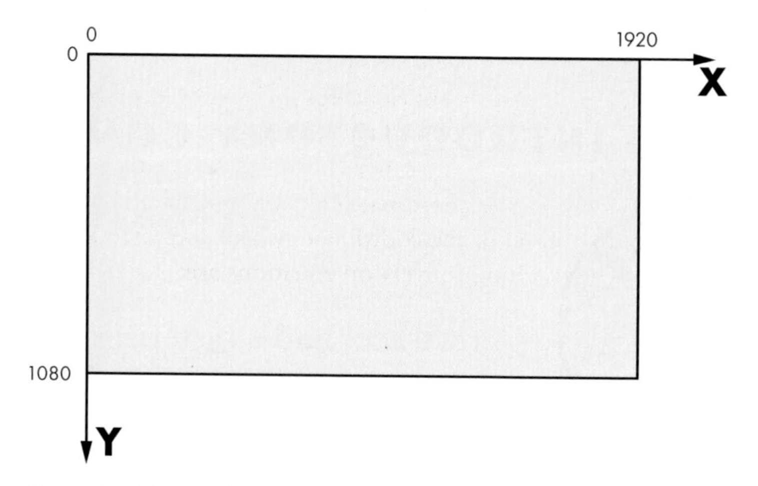

Figure 1-1: The coordinate system used by most computer screens

This coordinate system is very natural for a computer because of the way video memory is organized, but it's not the most natural for humans to work with. Instead, 3D graphics programmers tend to use the coordinate system typically used to draw graphs on paper: the origin is at the center, *x* increases toward the right and decreases toward the left, while *y* increases toward the top and decreases toward the bottom, as in Figure 1-2.

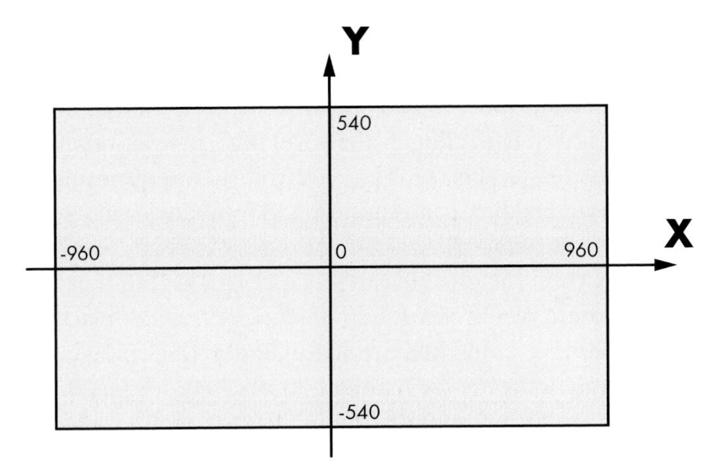

Figure 1-2: The coordinate system we'll use for our canvas

Using this coordinate system, the range of the *x* coordinate is $[\frac{-C_w}{2}, \frac{C_w}{2})$ and the range of the *y* coordinate is $[\frac{-C_h}{2}, \frac{C_h}{2})$. Let's assume that using the `PutPixel` function with coordinates outside these ranges does nothing.

In our examples, the canvas will be drawn on the screen, so we'll need to convert from one coordinate system to the other. To do this, we need to change the center of the coordinate system and reverse the direction of the *Y* axis. The resulting conversion equations are:

$$S_x = \frac{C_w}{2} + C_x$$

$$S_y = \frac{C_h}{2} - C_y$$

We will assume `PutPixel` does this conversion automatically; from this point on, we can think of the canvas as having its coordinate origin at the center, with *x* increasing to the right and *y* increasing to the top of the screen.

Let's take a look at the remaining argument of `PutPixel`: the color.

Color Models

The theory of how color works is fascinating, but it's outside the scope of this book. The following is a simplified version of the aspects relevant to us.

When light hits our eyes, it stimulates the light-sensitive cells at the back of them. These generate brain signals that depend on the wavelength of the incoming light. We call the subjective experience of these brain signals *colors*.

We can't normally see wavelengths outside of the *visible range*. Wavelength and frequency are inversely related (the more frequently the wave hits, the smaller the distance between the peaks of that wave). This is why infrared (wavelengths longer than 740 nm, corresponding to frequencies lower than 405 terahertz [THz]) is harmless, but ultraviolet (wavelengths shorter than 380 nm, corresponding to frequencies higher than 790 THz) can burn your skin.

Every color imaginable can be described as different combinations of these colors. "White" is the sum of all colors, while "black" is the absence of any colors. It would be impractical to describe colors by describing the exact wavelengths they're made of; fortunately, it's possible to create almost all colors as a linear combination of just three colors, which we call *primary colors*.

Subtractive Color Model

The *subtractive color model* is a fancy name for that thing you did with crayons as a toddler. You take a white piece of paper and red, blue, and yellow crayons. You draw a yellow circle, then a blue circle that overlaps it, and you get green! Yellow and red—orange! Red and blue—purple! Mix the three together—something darkish! Wasn't kindergarten amazing? Figure 1-3 shows the primary colors of the subtractive model, and the colors that result from mixing them.

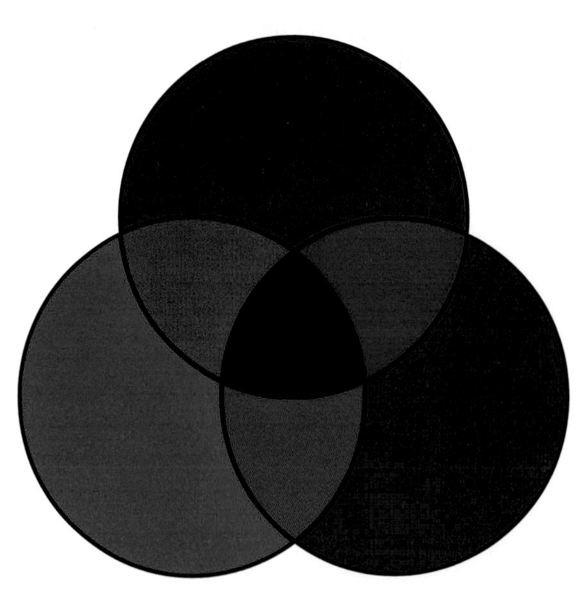

Figure 1-3: Subtractive primary colors and their combinations

Objects are of different colors because they absorb and reflect light in different ways. Let's start with white light, like sunlight (sunlight isn't quite white, but it's close enough for our purposes). White light contains light of every wavelength. When it hits an object, the object's surface absorbs some of the wavelengths and reflects others, depending on the material. Some of the reflected light then hits our eyes, and our brains convert that to color. What color? The sum of the wavelengths that were reflected by the surface.

So what's going on with the crayons? You start with white light reflecting off the paper. Since it's white paper, it reflects most of the light it receives. When you draw with a "yellow" crayon, you're adding a layer of a material that absorbs some wavelengths but lets others pass through it. They're reflected by the paper, pass through the yellow layer again, hit your eyes, and your brain interprets that particular combination of wavelengths as "yellow." What the yellow layer does is *subtract* a bunch of wavelengths from the original white light.

You can think of each colored circle as a filter: when you draw a blue circle overlapping the yellow one, you're filtering out even more wavelengths from the original light, so what hits your eyes is whatever wavelengths weren't filtered by either the blue or the yellow circles, which your brain sees as "green."

In summary, we start with all wavelengths and subtract some amount of the primary colors to create any other color. This color model gets its name from the fact that we're creating colors by subtracting wavelengths.

This model isn't quite right, though. The actual primary colors in the subtractive model aren't the blue, red, and yellow taught to toddlers and art students, but cyan, magenta, and yellow. Furthermore, mixing the three primary colors produces a somewhat darkish color that isn't quite black, so pure black is added as a fourth "primary." Because B is used to represent blue, black is denoted by K, and so we arrive at the *CMYK color model* (Figure 1-4).

You can see evidence of this color model directly on the cartridges of color printers, or sometimes in the shapes of cheaply printed flyers where the different colors are slightly offset from one another.

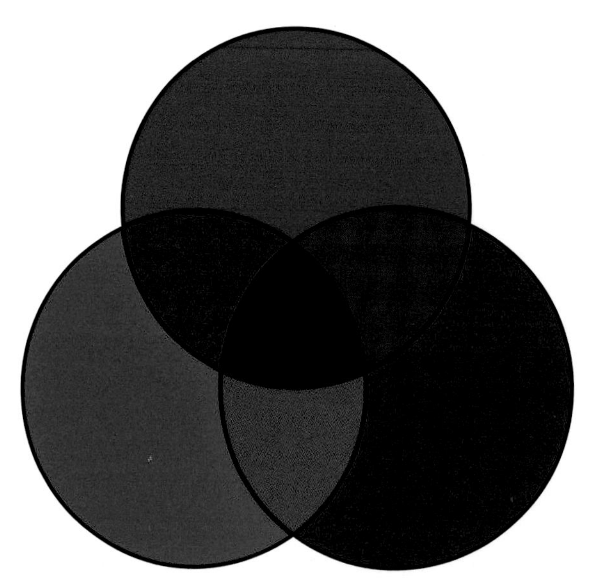

Figure 1-4: The four subtractive primary colors used by printers

Additive Color Model

The subtractive color model is only half the story. If you've ever looked at a screen up close or with a magnifying glass (or, let's be honest, accidentally sneezed on it), you've probably seen tiny colored dots: these are red, green, and blue.

Computer screens are the opposite of paper. Paper doesn't emit light; it merely reflects part of the light that hits it. Screens, on the other hand, are black, but they do emit light on their own. With paper, we start with white light and *subtract* the wavelengths we don't want; with a screen, we start with no light and *add* the wavelengths we want.

Different primary colors are necessary for this. Most colors can be created by adding different amounts of red, green, and blue to a black surface; this is the *RGB color model*, an *additive color model*, shown in Figure 1-5.

The combination of additive primary colors is *lighter* than its components, whereas the combination of subtractive primary colors is *darker*; all the additive primaries add up to white, while all the subtractive primaries add up to black.

Figure 1-5: The additive primary colors and some of their combinations

Forget the Details

Now that you know all this, you can selectively forget most of the details and focus on what's important for our work.

Most colors can be represented in either RGB or CMYK (or in any of the many other color models), and it's possible to convert from one *color space* to another. Since we're focusing on rendering things on a screen, we use the RGB color model for the rest of this book.

As described above, objects absorb part of the light reaching them and reflect the rest. Which wavelengths are absorbed and which are reflected is what we perceive as the "color" of the surface. From now on, we'll simply treat the color as a property of a surface and forget about light wavelengths.

Color Depth and Representation

Monitors create colors by mixing different amounts of red, green, and blue. They do this by lighting the tiny colored dots on their surface at different intensities by supplying different voltages to them.

How many different intensities can we get? Although voltage is continuous, we'll be manipulating colors with a computer, which uses discrete values (that is, a limited number of them). The more shades of red, green, and blue we can represent, the more colors we'll be able to produce.

Most images you see these days use 8 bits per primary color, which we call a *color channel* in this context. Using 8 bits per channel gives us 24 bits per pixel, for a total of 2^{24} different colors (approximately 16.7 million). This format, known as *R8G8B8* or simply *888*, is the one we'll use throughout this book. We say this format has a *color depth* of 24 bits.

This is by no means the only possible format. Not so long ago, in order to save memory, 15- and 16-bit formats were popular, assigning 5 bits per channel in the 15-bit case, and 5 bits for red, 6 for green, and 5 for blue in the 16-bit case (known as the *R5G6B5* or *565* format). Green gets the extra bit because our eyes are more sensitive to changes in green than to changes in red or blue.

With 16 bits, we get 2^{16} colors (approximately 65,000). This means you get one color for every 256 colors in 24-bit mode. Although 65,000 colors is plenty, for images where colors change very gradually you would be able to see very subtle "steps" that just aren't visible with 16.7 million colors, where there are enough bits to represent the colors in between. For some specialized applications, such as color grading for movies, it's a good idea to represent even more color detail, using even more bits per channel.

We'll use 3 bytes to represent a color, each holding the value of an 8-bit color channel, from 0 to 255. We'll express the colors as (R, G, B)—for example, $(255, 0, 0)$ represents pure red; $(255, 255, 255)$ represents white; and $(255, 0, 128)$ represents a reddish purple.

Color Manipulation

We'll use a handful of operations to manipulate colors. If you know some linear algebra, you can think of colors as vectors in 3D color space. If not, don't worry, we'll go through the basic operations we'll be using now.

We can modify the intensity of a color by multiplying each of its color channels by a constant:

$$k(R, G, B) = (kR, kG, kB)$$

For example, $(32, 0, 128)$ is twice as bright as $(16, 0, 64)$.

We can add two colors together by adding their color channels separately:

$$(R_1, G_1, B_1) + (R_2, G_2, B_2) = (R_1 + R_2, G_1 + G_2, B_1 + B_2)$$

For example, if we want to combine red $(255, 0, 0)$ and green $(0, 255, 0)$, we add them channel-wise and get $(255, 255, 0)$, which is yellow.

These operations can yield invalid values; for example, doubling the intensity of (192, 64, 32) produces an R value outside our color range. We'll treat any value over 255 as 255, and any value below 0 as 0; we call this *clamping* the value to the [0–255] range. This is more or less equivalent to what happens when you take an under- or over-exposed photograph in real life: you get either completely black or completely white areas.

That about sums it up for our primer on colors and `PutPixel`. Before we move on to the next chapter, let's spend a little time exploring how to represent the 3D objects we'll be rendering.

The Scene

So far, we have introduced the canvas, the abstract surface on which we can color pixels. Now we turn our attention to the objects we're interested in representing by introducing another abstraction: the *scene*.

The scene is the set of objects you may be interested in rendering. It could represent anything, from a single sphere floating in the empty infinity of space (we'll start there) to an incredibly detailed model of the inside of a grumpy ogre's nose.

We need a coordinate system to talk about objects within the scene. We can't use the same coordinate system as the canvas, for two reasons. First, the canvas is 2D, whereas the scene is 3D. Second, the canvas and the scene use different units: we use pixels for the canvas and real-world units (such as the imperial or metric systems) for the scene.

The choice of axes is arbitrary, so we'll pick something useful for our purposes. We'll say that Y is up and X and Z are horizontal, and all three axes are perpendicular to each other. Think of the plane XZ as the "floor," while XY and YZ are vertical "walls" in a square room. This is consistent with the coordinate system we chose for the canvas, where Y is up and X is horizontal. Figure 1-6 shows what this looks like.

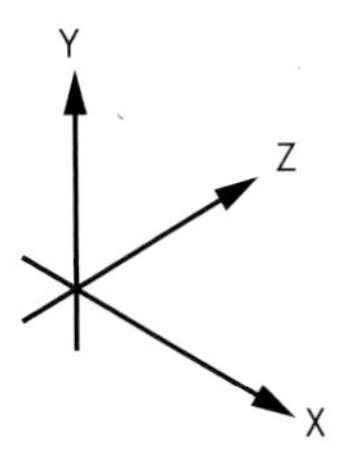

Figure 1-6: The coordinate system we'll use for our scenes

The choice of scene units is somewhat arbitrary; it depends on what your scene represents. A measurement of "1" could mean 1 inch if you're modeling a teacup, or it could mean 1 astronomical unit if you're modeling the Solar System. As long as we use our chosen units consistently, it doesn't matter what they are, so we can safely ignore them from now on.

Summary

In this chapter, we've introduced the canvas, an abstraction that represents a rectangular surface we can draw on, plus the one method we'll build everything else on: `PutPixel`. We've also chosen a coordinate system to refer to the pixels on the canvas and described a way to represent the color of these pixels. Lastly, we introduced the concept of a scene and chose a coordinate system to use in the scene.

Having laid these foundations, it's time to start building a raytracer and a rasterizer on top of them.

PART I

RAYTRACING

2

BASIC RAYTRACING

In this chapter, we'll introduce raytracing, the first major algorithm we'll cover. We start by motivating the algorithm and laying out some basic pseudocode. Then we look at how to represent rays of light and objects in a scene. Finally, we derive a way to compute which rays of light make up the visible image of each of the objects in our scene and see how we can represent them on the canvas.

Rendering a Swiss Landscape

Suppose you're visiting some exotic place and come across a stunning landscape—so stunning, you just *need* to make a painting capturing its beauty. Figure 2-1 shows one such landscape.

Figure 2-1: A breathtaking Swiss landscape

You have a canvas and a paint brush, but you absolutely lack artistic talent. Is all hope lost?

Not necessarily. You may not have artistic talent, but you are methodical. So you do the most obvious thing: you get an insect net. You cut a rectangular piece, frame it, and fix it to a stick. Now you can look at the landscape through a netted window. Next, you choose the best point of view to appreciate this landscape and plant another stick to mark the exact position where your eye should be.

You haven't started the painting yet, but now you have a fixed point of view and a fixed frame through which you can see the landscape. Moreover, this fixed frame is divided into small squares by the insect net. Now comes the methodical part. You draw a grid on the canvas, giving it the same number of squares as the insect net. Then you look at the top-left square of the net. What's the predominant color you can see through it? Sky blue. So you paint the top-left square of the canvas sky blue. You do this for every square, and soon enough the canvas contains a pretty good painting of the landscape, as seen through the frame. The resulting painting is shown in Figure 2-2.

Figure 2-2: A crude approximation of the landscape

When you think about it, a computer is essentially a very methodical machine absolutely lacking artistic talent. We can describe the process of creating our painting as follows:

```
For each little square on the canvas
    Paint it the right color
```

Easy! However, that formulation is too abstract to implement directly on a computer. We can go into a bit more detail:

```
Place the eye and the frame as desired
For each square on the canvas
    Determine which square on the grid corresponds to this square on the canvas
    Determine the color seen through that grid square
    Paint the square with that color
```

This is still too abstract, but it starts to look like an algorithm—and perhaps surprisingly, that's a high-level overview of the full raytracing algorithm! Yes, it's that simple.

Basic Assumptions

Part of the charm of computer graphics is drawing things on the screen. To achieve this as soon as possible, we'll make some simplifying assumptions.

Of course, these assumptions impose some restrictions over what we can do, but we'll lift the restrictions in later chapters.

First of all, we'll assume a fixed viewing position. The viewing position, the place where you'd put your eye in the Swiss landscape analogy, is commonly called the *camera position*; let's call it O. We'll assume that the camera occupies a single point in space, that it is located at the origin of the coordinate system, and that it never moves from there, so $O = (0, 0, 0)$ for now.

Second, we'll assume a fixed camera orientation. The camera orientation determines where the camera is pointing. We'll assume it looks in the direction of the positive Z axis (which we'll shorten to $\vec{Z_+}$), with the positive Y axis ($\vec{Y_+}$) up and the positive X axis ($\vec{X_+}$) to the right (Figure 2-3).

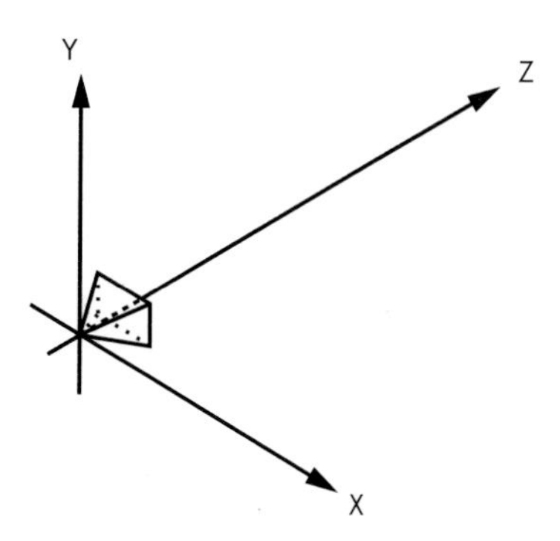

Figure 2-3: The position and orientation of the camera

The camera position and orientation are now fixed. Still missing from the analogy is the "frame" through which we look at the scene. We'll assume this frame has dimensions V_w and V_h, and is frontal to the camera orientation—that is, perpendicular to $\vec{Z_+}$. We'll also assume it's at a distance d, its sides are parallel to the X and Y axes, and it's centered with respect to $\vec{Z}$. That's a mouthful, but it's actually quite simple. Take a look at Figure 2-4.

The rectangle that will act as our window to the world is called the *viewport*. Essentially, we'll draw on the canvas whatever we see through the viewport. Note that the size of the viewport and the distance to the camera determine the angle visible from the camera, called the *field of view*, or FOV for short. Humans have an almost 180° horizontal FOV (although much of it is blurry peripheral vision with no sense of depth). For simplicity, we'll set $V_w = V_h = d = 1$; this results in a FOV of approximately 53°, which produces reasonable-looking images that are not overly distorted.

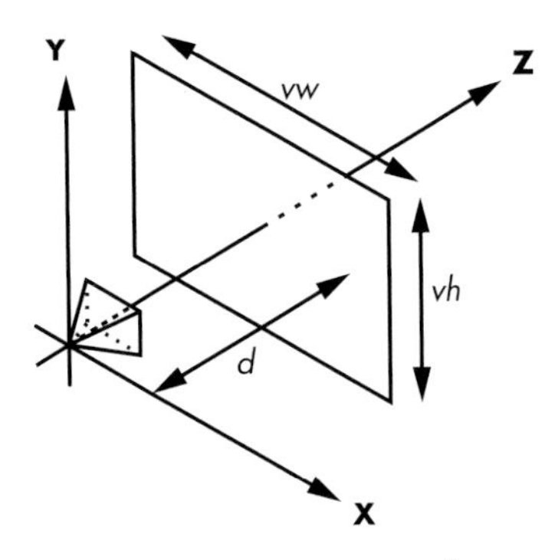

Figure 2-4: The position and orientation of the viewport

Let's go back to the "algorithm" presented earlier, use the appropriate technical terms, and number the steps in Listing 2-1.

```
❶ Place the camera and the viewport as desired
For each pixel on the canvas
    ❷ Determine which square on the viewport corresponds to this pixel
    ❸ Determine the color seen through that square
    ❹ Paint the pixel with that color
```

Listing 2-1: A high-level description of our raytracing algorithm

We have just done step ❶ (or, more precisely, gotten it out of the way for now). Step ❹ is trivial: we simply use `canvas.PutPixel(x, y, color)`. Let's do step ❷ quickly, and then focus our attention on increasingly sophisticated ways of doing step ❸ over the next few chapters.

Canvas to Viewport

Step ❷ of our algorithm in Listing 2-1 asks us to `Determine which square on the viewport corresponds to this pixel`. We know the canvas coordinates of the pixel—let's call them C_x and C_y. Notice how we conveniently placed the viewport so that its axes match the orientation of those of the canvas, and its center matches the center of the canvas. Because the viewport is measured in world units and the canvas is measured in pixels, going from canvas coordinates to space coordinates is just a change of scale!

$$V_x = C_x \cdot \frac{V_w}{C_w}$$

$$V_y = C_y \cdot \frac{V_h}{C_h}$$

There's an extra detail. Although the viewport is 2D, it's embedded in 3D space. We defined it to be at a distance d from the camera; every point in this plane (called the *projection plane*) has, by definition, $z = d$. Therefore,

$$V_z = d$$

And we're done with this step. For each pixel (C_x, C_y) on the canvas, we can determine its corresponding point on the viewport (V_x, V_y, V_z).

Tracing Rays

The next step is to figure out what color the light coming through (V_x, V_y, V_z) is, as seen from the camera's point of view (O_x, O_y, O_z).

In the real world, light comes from a light source (the Sun, a light bulb, and so on), bounces off several objects, and then finally reaches our eyes. We could try simulating the path of every photon leaving our simulated light sources, but it would be *extremely* time-consuming. Not only would we have to simulate a mind-boggling number of photons (a single 100 W light bulb emits 10^{20} photons per second!), only a tiny minority of them would happen to reach (O_x, O_y, O_z) after coming through the viewport. This technique is called *photon tracing* or *photon mapping*; unfortunately, it's outside the scope of this book.

Instead, we'll consider the rays of light "in reverse"; we'll start with a ray originating from the camera, going through a point in the viewport, and tracing its path until it hits some object in the scene. This object is what the camera "sees" through that point of the viewport. So, as a first approximation, we'll just take the color of that object as "the color of the light coming through that point," as shown in Figure 2-5.

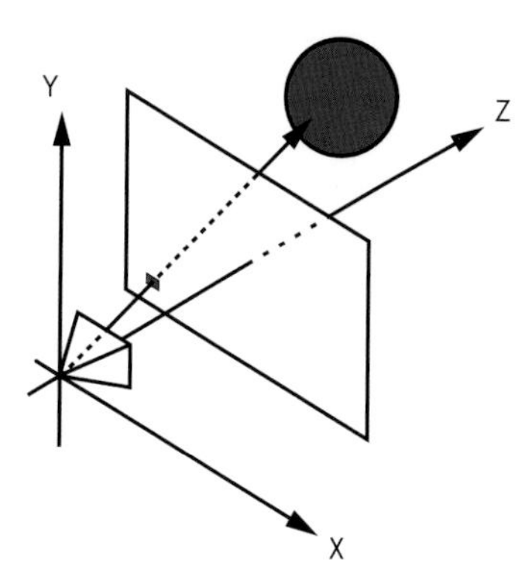

Figure 2-5: A tiny square in the viewport, representing a single pixel in the canvas, painted with the color of the object the camera sees through it

Now we just need some equations.

The Ray Equation

The most convenient way to represent a ray for our purposes is with a parametric equation. We know the ray passes through O, and we know its direction (from O to V), so we can express any point P in the ray as

$$P = O + t(V - O)$$

where t is any real number. By plugging every value of t from $-\infty$ to $+\infty$ into this equation, we get every point P along the ray.

Let's call $(V - O)$, the direction of the ray, $\vec{D}$. The equation becomes

$$P = O + t\vec{D}$$

An intuitive way to understand this equation is that we start the ray at the origin (O) and "advance" along the direction of the ray ($\vec{D}$) by some amount (t); it's easy to see that this includes all the points along the ray. You can read more details about these vector operations in the Linear Algebra appendix. Figure 2-6 shows our equation in action.

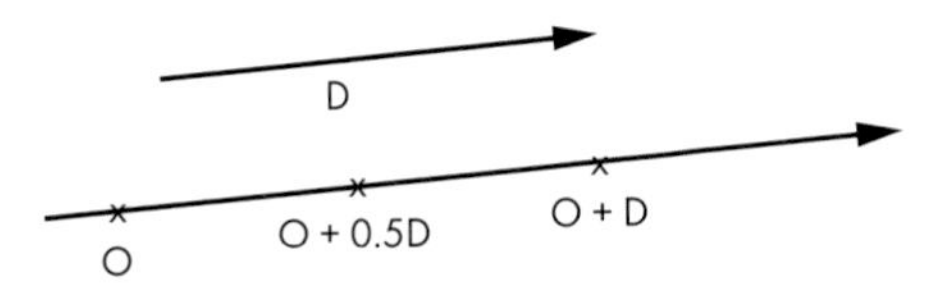

Figure 2-6: Some points of the ray $O + t\vec{D}$ *for different values of* t.

Figure 2-6 shows the points along the ray that corresponds to $t = 0.5$ and $t = 1.0$. Every value of t yields a different point along the ray.

The Sphere Equation

Now we need to have some sort of object in the scene, so that our rays can hit *something*. We could choose any arbitrary geometric primitive as the building block of our scenes; for raytracing, we'll use spheres because they're easy to manipulate with equations.

What is a sphere? A sphere is the set of points that lie at a fixed distance from a fixed point. That distance is called the *radius* of the sphere, and the point is called the *center* of the sphere. Figure 2-7 shows a sphere, defined by its center C and its radius r.

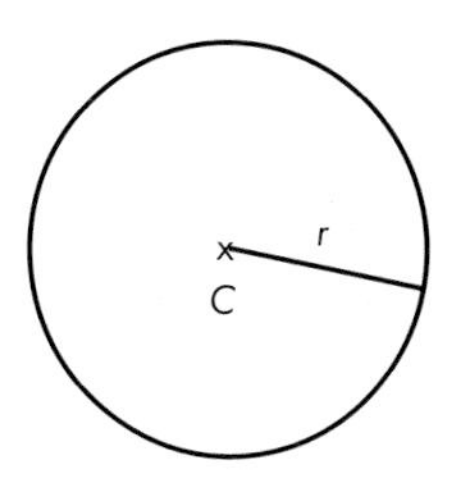

Figure 2-7: A sphere, defined by its center and its radius

According to our definition above, if C is the center and r is the radius of a sphere, the points P on the surface of that sphere must satisfy the following equation:

$$distance(P, C) = r$$

Let's play a bit with this equation. If you find any of this math unfamiliar, read through the Linear Algebra appendix.

The distance between P and C is the length of the vector from P to C:

$$|P - C| = r$$

The length of a vector (denoted $|\vec{V}|$) is the square root of its dot product with itself (denoted $\langle \vec{V}, \vec{V} \rangle$):

$$\sqrt{\langle P - C, P - C \rangle} = r$$

To get rid of the square root, we can square both sides:

$$\langle P - C, P - C \rangle = r^2$$

All these formulations of the sphere equation are equivalent, but this last one is the most convenient to manipulate in the following steps.

Ray Meets Sphere

We now have two equations: one describing the points on the sphere, and one describing the points on the ray:

$$\langle P - C, P - C \rangle = r^2$$

$$P = O + t\vec{D}$$

Do the ray and the sphere intersect? If so, where?

Suppose the ray and the sphere do intersect at a point P. This point is both along the ray and on the surface of the sphere, so it must satisfy both equations at the same time. Note that the only variable in these equations is the parameter t, since O, $\vec{D}$, C, and r are given and P is the point we're trying to find.

Since P represents the same point in both equations, we can substitute P in the first one with the expression for P in the second. This gives us

$$\langle O + t\vec{D} - C, O + t\vec{D} - C \rangle = r^2$$

If we can find values of t that satisfy this equation, we can put them in the ray equation to find the points where the ray intersects the sphere.

In its current form, the equation is somewhat unwieldy. Let's do some algebraic manipulation to see what we can get out of it.

First, let $\vec{CO} = O - C$. Then we can write the equation as

$$\langle \vec{CO} + t\vec{D}, \vec{CO} + t\vec{D} \rangle = r^2$$

Then we expand the dot product into its components, using its distributive properties (again, feel free to consult the Linear Algebra appendix):

$$\langle \vec{CO} + t\vec{D}, \vec{CO} \rangle + \langle \vec{CO} + t\vec{D}, t\vec{D} \rangle = r^2$$

$$\langle \vec{CO}, \vec{CO} \rangle + \langle t\vec{D}, \vec{CO} \rangle + \langle \vec{CO}, t\vec{D} \rangle + \langle t\vec{D}, t\vec{D} \rangle = r^2$$

Rearranging the terms a bit, we get

$$\langle t\vec{D}, t\vec{D} \rangle + 2\langle \vec{CO}, t\vec{D} \rangle + \langle \vec{CO}, \vec{CO} \rangle = r^2$$

Moving the parameter t out of the dot products and moving r^2 to the other side of the equation gives us

$$t^2 \langle \vec{D}, \vec{D} \rangle + t(2\langle \vec{CO}, \vec{D} \rangle) + \langle \vec{CO}, \vec{CO} \rangle - r^2 = 0$$

Remember that the dot product of two vectors is a real number, so every term between angle brackets is a real number. If we give them names, we'll get something much more familiar:

$$a = \langle \vec{D}, \vec{D} \rangle$$

$$b = 2\langle \vec{CO}, \vec{D} \rangle$$

$$c = \langle \vec{CO}, \vec{CO} \rangle - r^2$$

$$at^2 + bt + c = 0$$

This is nothing more and nothing less than a good old quadratic equation! Its solutions are the values of the parameter t where the ray intersects the sphere:

$$\{t_1, t_2\} = \frac{-b \pm \sqrt{b^2 - 4ac}}{2a}$$

Fortunately, this makes geometrical sense. As you may remember, a quadratic equation can have no solutions, one double solution, or two different solutions, depending on the value of the discriminant $b^2 - 4ac$. This corresponds exactly to the cases where the ray doesn't intersect the sphere, the ray is tangent to the sphere, and the ray enters and exits the sphere, respectively (Figure 2-8).

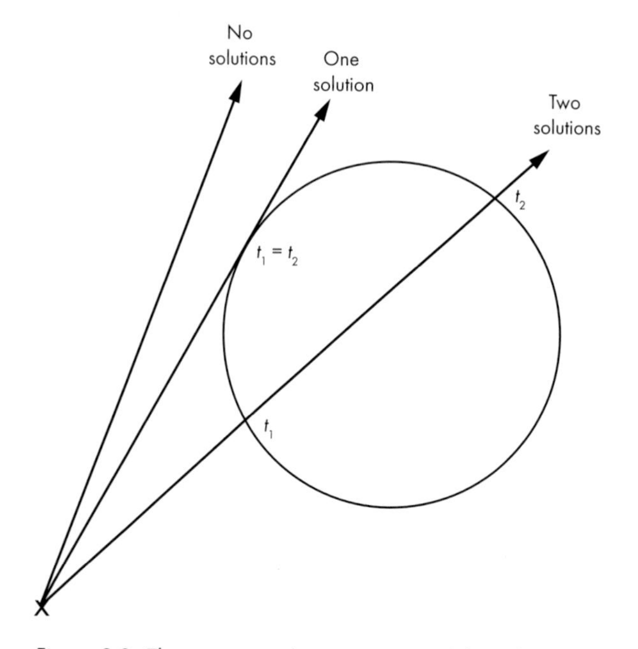

Figure 2-8: The geometrical interpretation of the solutions to the quadratic equation: no solutions, one solution, or two solutions.

Once we have found the value of t, we can plug it back into the ray equation, and we finally get the intersection point P corresponding to that value of t.

Rendering our First Spheres

To recap, for each pixel on the canvas, we can compute the corresponding point on the viewport. Given the position of the camera, we can express the equation of a ray that starts at the camera and goes through that point of the viewport. Given a sphere, we can compute where the ray intersects that sphere.

So all we need to do is to compute the intersections of the ray and each sphere, keep the intersection closest to the camera, and paint the pixel on the canvas with the appropriate color. We're almost ready to render our first spheres!

The parameter t deserves some extra attention, though. Let's go back to the ray equation:

$$P = O + t(V - O)$$

Since the origin and direction of the ray are fixed, varying t across all the real numbers will yield every point P in this ray. Note that for $t = 0$ we get $P = O$, and for $t = 1$ we get $P = V$. Negative values of t yield points in the *opposite* direction—that is, *behind* the camera. So, we can divide the parameter space into three parts, as in Table 2-1. Figure 2-9 shows a diagram of the parameter space.

Table 2-1: Subdivisions of the Parameter Space

$t < 0$	Behind the camera
$0 \leq t \leq 1$	Between the camera and the projection plane/viewport
$t > 1$	In front of the projection plane/viewport

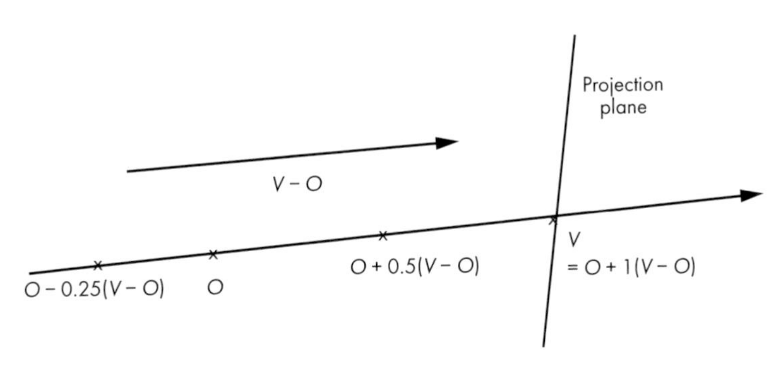

Figure 2-9: A few points in parameter space

Note that nothing in the intersection equation says that the sphere has to be *in front* of the camera; the equation will happily produce solutions for intersections *behind* the camera. Obviously, this isn't what we want, so we should ignore any solutions with $t < 0$. To avoid further mathematical unpleasantness, we'll restrict the solutions to $t > 1$; that is, we'll render whatever is *beyond* the projection plane.

On the other hand, we don't want to put an upper bound on the value of t; we want to see all objects in front of the camera, no matter how far away they are. However, because in later stages we *will* want to cut rays short, we'll introduce this formalism now and give t an upper value of $+\infty$ (for languages that can't represent "infinity" directly, a really really big number does the trick).

We can now formalize everything we've done so far with some pseudocode. As a general rule, we'll assume the code has access to whatever data it needs, so we won't bother explicitly passing around parameters such as the canvas and will focus on the really necessary ones.

The main method now looks like Listing 2-2.

```
O = (0, 0, 0)
for x = -Cw/2 to Cw/2 {
    for y = -Ch/2 to Ch/2 {
        D = CanvasToViewport(x, y)
        color = TraceRay(O, D, 1, inf)
        canvas.PutPixel(x, y, color)
    }
}
```

Listing 2-2: The main method

The `CanvasToViewport` function is very simple, and is shown in Listing 2-3. The constant `d` represents the distance between the camera and the projection plane.

```
CanvasToViewport(x, y) {
    return (x*Vw/Cw, y*Vh/Ch, d)
}
```

Listing 2-3: The `CanvasToViewport` function

The `TraceRay` method (Listing 2-4) computes the intersection of the ray with every sphere and returns the color of the sphere at the nearest intersection inside the requested range of t.

```
TraceRay(O, D, t_min, t_max) {
    closest_t = inf
    closest_sphere = NULL
    for sphere in scene.spheres {
        t1, t2 = IntersectRaySphere(O, D, sphere)
        if t1 in [t_min, t_max] and t1 < closest_t {
            closest_t = t1
            closest_sphere = sphere
        }
        if t2 in [t_min, t_max] and t2 < closest_t {
            closest_t = t2
            closest_sphere = sphere
        }
```

```
    }
    if closest_sphere == NULL {
      ❶ return BACKGROUND_COLOR
    }
    return closest_sphere.color
}
```

Listing 2-4: The `TraceRay` method

In Listing 2-4, `O` represents the origin of the ray; although we're tracing rays from the camera, which is placed at the origin, this won't necessarily be the case in later stages, so it has to be a parameter. The same applies to `t_min` and `t_max`.

Note that when the ray doesn't intersect any sphere, we still need to return *some* color ❶—I've chosen white in most of these examples.

Finally, `IntersectRaySphere` (Listing 2-5) just solves the quadratic equation.

```
IntersectRaySphere(O, D, sphere) {
    r = sphere.radius
    CO = O - sphere.center

    a = dot(D, D)
    b = 2*dot(CO, D)
    c = dot(CO, CO) - r*r

    discriminant = b*b - 4*a*c
    if discriminant < 0 {
        return inf, inf
    }

    t1 = (-b + sqrt(discriminant)) / (2*a)
    t2 = (-b - sqrt(discriminant)) / (2*a)
    return t1, t2
}
```

Listing 2-5: The `IntersectRaySphere` method

To put all of this into practice, let's define a very simple scene, shown in Figure 2-10.

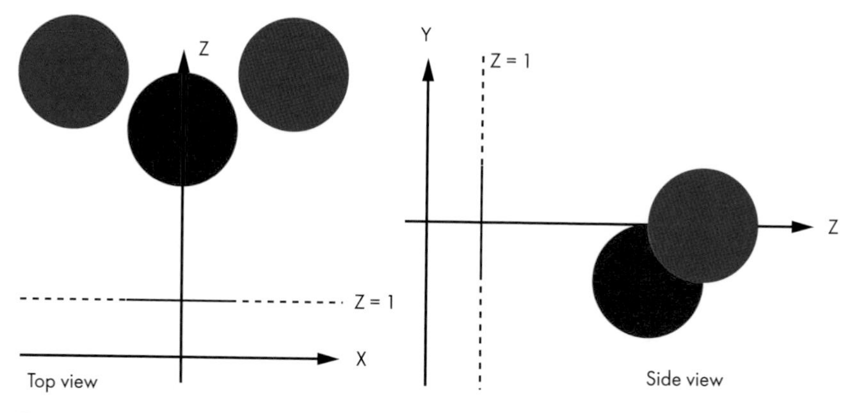

Figure 2-10: A very simple scene, viewed from above (left) and from the right (right)

In pseudoscene language, it's something like this:

```
viewport_size = 1 x 1
projection_plane_d = 1
sphere {
    center = (0, -1, 3)
    radius = 1
    color = (255, 0, 0)  # Red
}
sphere {
    center = (2, 0, 4)
    radius = 1
    color = (0, 0, 255)  # Blue
}
sphere {
    center = (-2, 0, 4)
    radius = 1
    color = (0, 255, 0)  # Green
}
```

When we run our algorithm on this scene, we're finally rewarded with an incredibly awesome raytraced scene (Figure 2-11).

Figure 2-11: An incredibly awesome raytraced scene

You can find a live implementation of this algorithm at *https://gabrielgambetta.com/cgfs/basic-rays-demo*.

I know, it's a bit of a letdown, isn't it? Where are the reflections and the shadows and the polished look? Don't worry, we'll get there. This is a good first step. The spheres look like circles, which is better than if they looked like cats. The reason they don't look quite like spheres is that we're missing a key component of how human beings determine the shape of an object: the way it interacts with light. We'll cover that in the next chapter.

Summary

In this chapter, we've laid down the foundations of our raytracer. We've chosen a fixed setup (the position and orientation of the camera and the viewport, as well as the size of the viewport); we've chosen representations for spheres and rays; we've explored the math necessary to figure out how spheres and rays interact; and we've put all this together to draw the spheres on the canvas using solid colors.

The next chapters build on this by modeling the way the rays of light interact with objects in the scene in increasing detail.

3

LIGHT

We'll start adding "realism" to our rendering of the scene by introducing light. Light is a vast and complex topic, so we'll present a simplified model that is good enough for our purposes. This model is, for the most part, inspired by how light works in the real world, but it also takes some liberties with the aim of making the rendered scenes look good.

We'll start with some simplifying assumptions that will make our lives easier, then we'll introduce three types of light sources: point lights, directional lights, and ambient light. We'll end the chapter by discussing how these lights affect the appearance of surfaces, including diffuse and specular reflection.

Simplifying Assumptions

Let's make a few assumptions to make things simpler. First, we declare that all light is white. This lets us characterize any light using a single real number,

i, called the *intensity* of the light. Simulating colored lights isn't that complicated (we'd just use three intensity values, one per color channel, and compute all color and lighting channel-wise), but we'll stick to white lights to keep things simple.

Second, we'll ignore the atmosphere. In real life, lights look dimmer the farther away they are; this is because of particles floating in the air that absorb part of the light as it travels through them. While this isn't particularly complicated to do in a raytracer, we'll keep it simple and ignore this effect; in our scene, distance doesn't make lights any less bright.

Light Sources

Light has to come from somewhere. In this section, we'll define three different types of light sources.

Point Lights

Point lights emit light from a fixed point in 3D space, called their *position*. They emit light equally in every direction; this is why they are also called *omnidirectional lights*. A point light is therefore fully described by its position and its intensity.

A light bulb is a good real-life approximation of a point light. While a real-life light bulb doesn't emit light from a single point, and it isn't perfectly omnidirectional, it's a pretty accurate approximation.

Let's define the vector $\vec{L}$ as the direction from a point in the scene, P, to the light, Q. We can calculate this vector, called the *light vector*, as $Q - P$. Note that since Q is fixed but P can be any point in the scene, $\vec{L}$ is different for every point in the scene, as you can see in Figure 3-1.

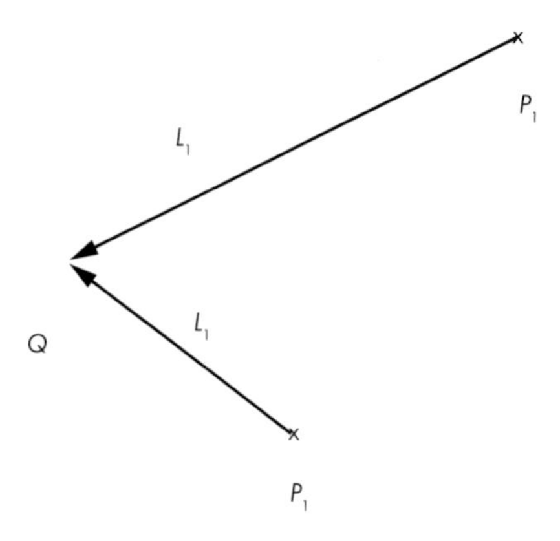

Figure 3-1: A point light at Q. The $\vec{L}$ vector is different for every point P.

Directional Lights

If a point light is a good approximation of a light bulb, does it also work as an approximation of the Sun?

This is a tricky question, and the answer depends on what we are trying to render. At the solar-system scale, the Sun can be approximated as a point light. After all, it emits light from a point, and it emits in all directions, so it seems to qualify.

However, if our scene represents something happening on Earth, it's not such a good approximation. The Sun is so far away that every ray of light that reaches us has almost exactly the same direction. We could approximate this with a point light located very, very, very far away from the objects in the scene. However, the distance between the light and the objects would be orders of magnitude greater than the distance between objects, so we'd start running into numerical accuracy errors.

To better handle these situations, we define *directional lights*. Like point lights, directional lights have an intensity, but unlike them, they don't have a position; instead, they have a fixed *direction*. You can think of them as infinitely distant point lights located in the specified direction.

While in the case of point lights we need to compute a different light vector $\vec{L}$ for every point P in the scene, in this case $\vec{L}$ is given. In the Sun-to-Earth scene example, $\vec{L}$ would be *(center of Sun) − (center of Earth)*. Figure 3-2 shows what this looks like.

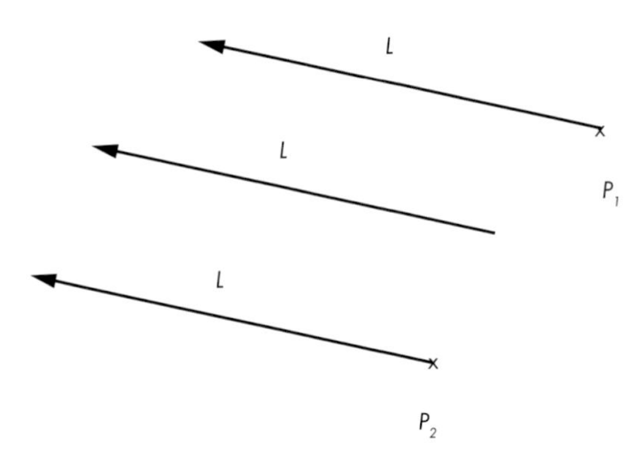

Figure 3-2: A directional light. The $\vec{L}$ *vector is the same for every point* P*.*

As we can see here, the light vector of a directional light is the same for every point in the scene. Compare this with Figure 3-1, where the light vector of a point light is different for every point in the scene.

Ambient Light

Can every real-life light be modeled as a point or directional light? Pretty much. Are these two types of light enough to light a scene? Unfortunately not.

Consider what happens to the Moon. The only significant light source nearby is the Sun. So the "front half" of the Moon with respect to the Sun gets all its light, and its "back half" is in complete darkness. We see this from different angles from Earth, creating what we call the "phases" of the Moon.

However, the situation down here on Earth is a bit different. Even points that don't receive light directly from a light source aren't completely in the dark (just look at the floor under your chair). How do rays of light reach these points if their "view" of the light sources is obstructed by something else?

As mentioned in "Color Models" in Chapter 1, when light hits an object, part of it is absorbed, but the rest is scattered back into the scene. This means that light can come not only from light sources, but also from objects that get light from light sources and scatter part of it back into the scene. But why stop there? The scattered light will in turn hit some other object, part of it will be absorbed, and part of it will be scattered back into the scene. And so on, until all of the energy of the original light has been absorbed by the surfaces in the scene.

This means we should treat *every object* as a light source. As you can imagine, this would add a lot of complexity to our model, so we won't explore that mechanism in this book. If you're curious, search for *global illumination* and marvel at the pretty pictures.

But we still don't want every object to be either directly illuminated or completely dark (unless we're actually rendering a model of the solar system). To overcome this limitation, we'll define a third type of light source, called *ambient light*, which is characterized only by its intensity. We'll declare that an ambient light contributes some light to every point in the scene, regardless of where it is. It's a gross oversimplification of the very complex interaction between the light sources and the surfaces in the scene, but it works well enough.

In general, a scene will have a single ambient light (because ambient lights only have an intensity value, any number of them can be trivially combined into a single ambient light) and an arbitrary number of point and directional lights.

Illumination of a Single Point

Now that we know how to define the lights in a scene, we need to figure out how the lights interact with the surfaces of the objects in the scene.

In order to compute the illumination of a single point, we'll compute the amount of light contributed by each light source and add them together to get a single number representing the total amount of light the point receives. We can then multiply the color of the surface at that point by this amount to get the shade of color that represents how much light it receives.

So, what happens when a ray of light, be it from a directional light or a point light, hits a point on some object in our scene?

We can intuitively classify objects into two broad classes, depending on how they reflect light: "matte" and "shiny" objects. Since most objects around us can be classified as matte, we'll focus on this group first.

Diffuse Reflection

When a ray of light hits a matte object, the ray is scattered back into the scene equally in every direction, a process called *diffuse reflection*; this is what makes matte objects look matte.

To verify this, look at some matte object around you, such as a wall. If you move with respect to the wall, its color doesn't change. That is, the light you see reflected from the object is the same no matter where you're looking from.

On the other hand, the amount of light reflected does depend on the *angle* between the ray of light and the surface. Intuitively, this happens because the energy carried by the ray has to spread over a smaller or bigger area depending on the angle, so the energy reflected to the scene per unit of area is higher or lower, respectively, as shown in Figure 3-3.

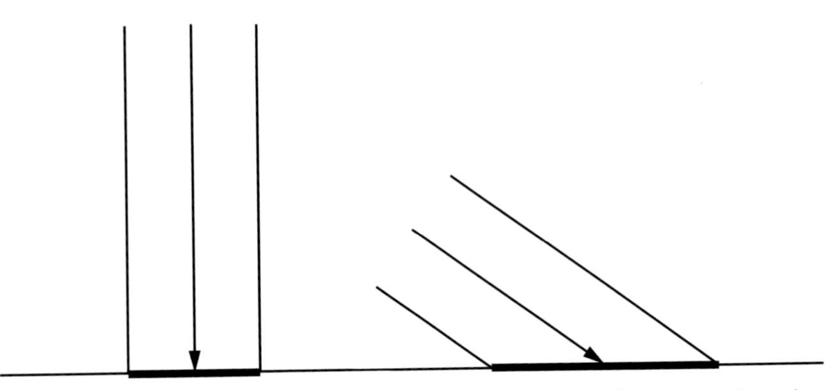

Figure 3-3: The energy of a ray of light spreads over areas of different size, depending on its angle to the surface.

In Figure 3-3, we can see two rays of light of the same intensity (represented by having the same width) hitting a surface head-on and at an angle. The energy carried by the rays of light spreads uniformly across the areas they hit. The energy of the ray on the right spreads across a bigger area than that of the ray on the left, and therefore each point in its area receives less energy than in the left-hand case.

To explore this mathematically, let's characterize the orientation of a surface by its *normal vector*. The normal vector of a surface at point P, or simply "the normal," is a vector perpendicular to the surface at P. It's also a unit vector, meaning its length is 1. We'll call this vector $\vec{N}$.

Modeling Diffuse Reflection

A ray of light with direction $\vec{L}$ and intensity I hits a surface with normal $\vec{N}$. What fraction of I is reflected back to the scene, as a function of I, $\vec{N}$, and $\vec{L}$?

As a geometric analogy, let's represent the intensity of the light as the "width" of the ray. Its energy spreads over a surface of size A. When $\vec{N}$ and $\vec{L}$ have the same direction–when the ray is perpendicular to the surface–then $I = A$, which means the energy reflected per unit of area is the same as the incident energy per unit of area: $\frac{I}{A} = 1$. On the other hand, as the angle between $\vec{L}$ and $\vec{N}$ approaches 90°, A approaches ∞, so the energy per unit of area approaches 0; $\lim_{A \to \infty} \frac{I}{A} = 0$. But what happens in between?

The situation is depicted in Figure 3-4. We know $\vec{N}$, $\vec{L}$, and P; I have added the angles α and β, and the points Q, R, and S to make writing about the diagram easier.

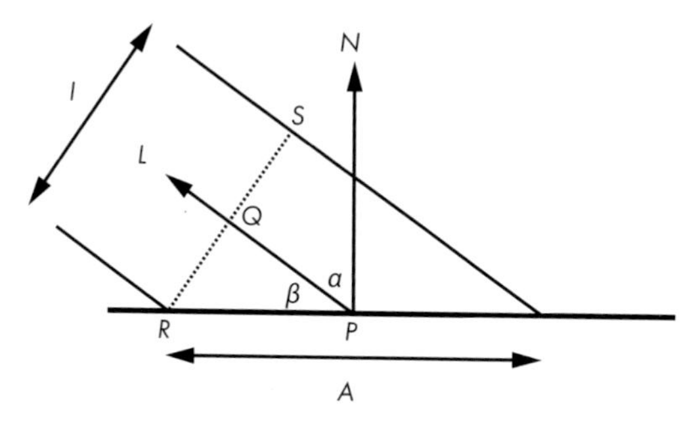

Figure 3-4: The vectors and angles involved in the diffuse reflection calculations

Since a ray of light technically has no width, we can assume that everything happens in a flat, infinitesimally small patch of the surface. Even if it's the surface of a sphere, the area we're considering is so infinitesimally small that it's almost flat in comparison with the size of the sphere, just like Earth looks flat at small scales.

The ray of light, with a width of I, hits the surface at P, at an angle β. The normal at P is $\vec{N}$, and the energy carried by the ray spreads over A. We need to compute $\frac{I}{A}$.

Consider RS, the "width" of the ray. By definition, it's perpendicular to $\vec{L}$, which is also the direction of PQ. Therefore, PQ and QR form a right angle, making PQR a right triangle.

One of the angles of PQR is $90°$, and another is β. The remaining angle is therefore $90° - \beta$. But note that $\vec{N}$ and PR also form a right angle, which means $\alpha + \beta$ must also be $90°$. Therefore, $\widehat{QRP} = \alpha$.

Let's focus on the triangle PQR (Figure 3-5). Its angles are α, β, and $90°$. The side QR measures $\frac{I}{2}$, and the side PR measures $\frac{A}{2}$.

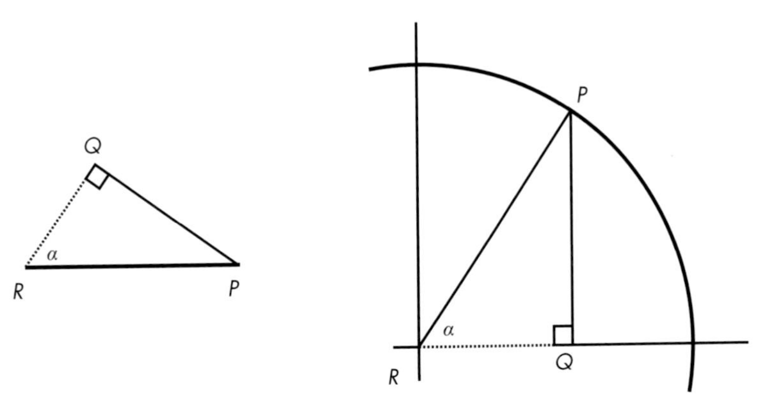

Figure 3-5: The PQR *triangle in a trigonometry context*

And now, trigonometry to the rescue! By definition, $\cos(\alpha) = \frac{QR}{PR}$; substituting QR with $\frac{I}{2}$ and PR with $\frac{A}{2}$, we get

$$\cos(\alpha) = \frac{\frac{I}{2}}{\frac{A}{2}}$$

which becomes

$$\cos(\alpha) = \frac{I}{A}$$

We're almost there. α is the angle between $\vec{N}$ and $\vec{L}$. We can use the properties of the dot product (feel free to consult the Linear Algebra appendix) to express $\cos(\alpha)$ as

$$\cos(\alpha) = \frac{\langle \vec{N}, \vec{L} \rangle}{|\vec{N}||\vec{L}|}$$

And finally

$$\frac{I}{A} = \frac{\langle \vec{N}, \vec{L} \rangle}{|\vec{N}||\vec{L}|}$$

We have arrived at a simple equation that gives us the fraction of light that is reflected as a function of the angle between the surface normal and the direction of the light.

Note that the value of $\cos(\alpha)$ becomes negative for angles over $90°$. If we blindly use this value, we can end up with a light source that makes a surface *darker*! This doesn't make any physical sense; an angle over 90° just means the light is actually illuminating the *back* of the surface, and therefore it doesn't contribute any light to the point we're illuminating. So if $\cos(\alpha)$ becomes negative, we need to treat it as if it was 0.

The Diffuse Reflection Equation

We can now formulate an equation to compute the full amount of light received by a point P with normal $\vec{N}$ in a scene with an ambient light of intensity I_A and n point or directional lights with intensity I_n and light vectors $\vec{L_n}$ either known (for directional lights) or computed for P (for point lights):

$$I_P = I_A + \sum_{i=1}^{n} I_i \frac{\langle \vec{N}, \vec{L_i} \rangle}{|\vec{N}|\,|\vec{L_i}|}$$

It's worth repeating that the terms where $\langle \vec{N}, \vec{L_i} \rangle < 0$ shouldn't be added to the point's illumination.

Sphere Normals

There's only a small detail missing: where do the normals come from? The answer to this general question is far more complex than it might seem, as we'll see in the second part of this book. Fortunately, at this point we're only dealing with spheres, and there's a very simple answer for them: the normal vector of any point of a sphere lies on a line that goes through the center of the sphere. As you can see in Figure 3-6, if the sphere center is C, the direction of the normal at point P is $P - C$.

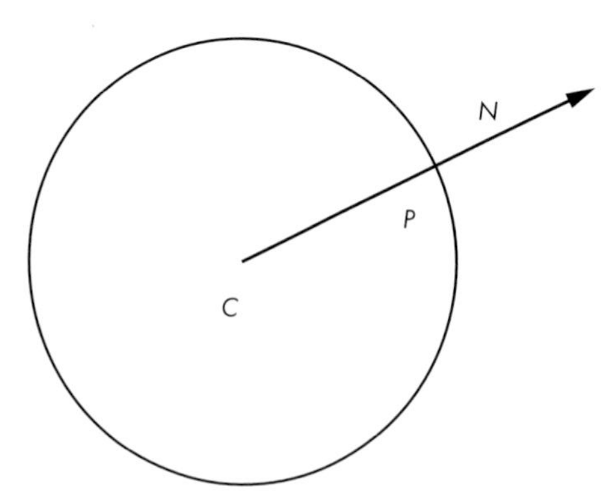

Figure 3-6: The normal of a sphere at P has the same direction as CP.

Why "the direction of the normal" and not "the normal"? A normal vector needs to be perpendicular to the surface, but it also needs to have length 1. To *normalize* this vector and turn it into a true normal, we need to divide it by its own length, thus guaranteeing the result has length 1:

$$\vec{N} = \frac{P - C}{|P - C|}$$

Rendering with Diffuse Reflection

Let's translate all of this to pseudocode. First, let's add a couple of lights to the scene:

```
light {
    type = ambient
    intensity = 0.2
}
light {
    type = point
    intensity = 0.6
    position = (2, 1, 0)
}
light {
    type = directional
    intensity = 0.2
    direction = (1, 4, 4)
}
```

Note that the intensities conveniently add up to 1.0; because of the way the lighting equation works, this ensures that no point can have a light intensity greater than this value. This means we won't have any "overexposed" spots.

The lighting equation is fairly straightforward to translate to pseudocode (Listing 3-1).

```
ComputeLighting(P, N) {
    i = 0.0
    for light in scene.Lights {
        if light.type == ambient {
          ❶ i += light.intensity
        } else {
            if light.type == point {
              ❷ L = light.position - P
            } else {
              ❸ L = light.direction
            }
```

```
            n_dot_l = dot(N, L)
        ❹ if n_dot_l > 0 {
            ❺ i += light.intensity * n_dot_l/(length(N) * length(L))
            }
        }
    }
    return i
}
```

Listing 3-1: A function to compute lighting with diffuse reflection

In Listing 3-1, we treat the three types of light in slightly different ways. Ambient lights are the simplest and are handled directly ❶. Point and directional lights share most of the code, in particular the intensity calculation ❺, but the direction vectors are computed in different ways (❷ and ❸), depending on their type. The condition in ❹ makes sure we don't add negative values, which represent lights illuminating the back side of the surface, as we discussed before.

The only thing left to do is to use `ComputeLighting` in `TraceRay`. We replace the line that returns the color of the sphere:

```
    return closest_sphere.color
```

with this snippet:

```
    P = O + closest_t * D  // Compute intersection
    N = P - closest_sphere.center  // Compute sphere normal at intersection
    N = N / length(N)
    return closest_sphere.color * ComputeLighting(P, N)
```

Just for fun, let's add a big yellow sphere:

```
sphere {
    color = (255, 255, 0)  # Yellow
    center = (0, -5001, 0)
    radius = 5000
}
```

We run the renderer and, lo and behold, the spheres now start to look like spheres (Figure 3-7)!

Figure 3-7: Diffuse reflection adds a sense of depth and volume to the scene.

You can find a live implementation of this algorithm at *https://gabrielgambetta.com/cgfs/diffuse-demo*.

But wait, how did the big yellow sphere turn into a flat yellow floor? It hasn't; it's just so big compared to the other three spheres, and the camera is so close to it, that it looks flat—just like the surface of our planet looks flat when we're standing on it.

Specular Reflection

Let's turn our attention to *shiny* objects. Unlike matte objects, shiny objects look slightly different depending on where you're looking from.

Imagine a billiard ball or a car just out of the car wash. These kinds of objects exhibit very specific light patterns, usually bright spots, that seem to move as you move around them. Unlike matte objects, the way you perceive the surface of these objects does actually depend on your point of view.

Note that a red billiard ball stays red if you walk around it, but the bright white spot that gives it its shiny appearance moves as you do. This shows that the new effect we want to model doesn't replace diffuse reflection, but instead complements it.

To understand why this happens, let's take a closer look at how surfaces reflect light. As we saw in the previous section, when a ray of light hits the surface of a matte object, it's scattered back to the scene equally in every direction. This happens because the surface of the object is irregular, so at the microscopic level it behaves like a set of tiny surfaces pointing in random directions (Figure 3-8).

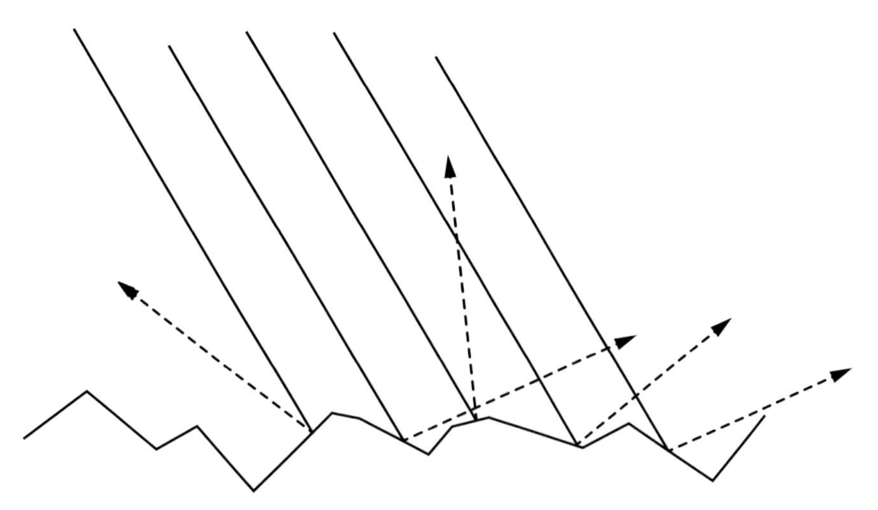

Figure 3-8: What the rough surface of a matte object might look like through a microscope. The incident rays of light are reflected in random directions.

But what if the surface isn't that irregular? Let's go to the other extreme: a perfectly polished mirror. When a ray of light hits a mirror, it's reflected in a single direction. If we call the direction of the reflected light $\vec{R}$, and we keep the convention that $\vec{L}$ points toward the light source, Figure 3-9 illustrates the situation.

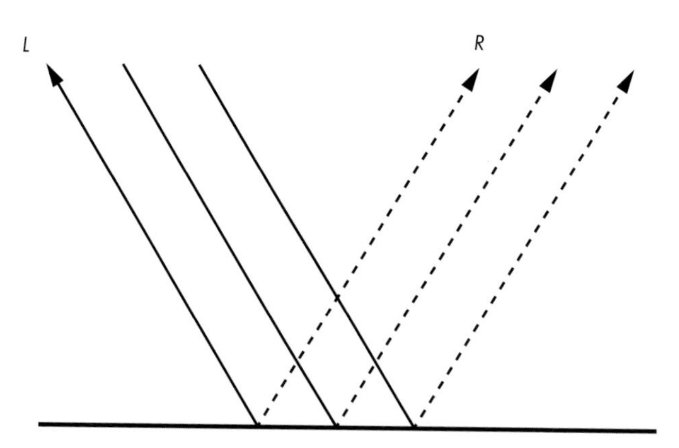

Figure 3-9: Rays of light reflected by a mirror

Depending on how "polished" the surface is, it behaves more or less like a mirror; this is why it's called *specular* reflection, from *speculum*, the Latin word for *mirror*.

For a perfectly polished mirror, the incident ray of light $\vec{L}$ is reflected in a single direction, $\vec{R}$. This is why you see reflected objects very clearly: for every incident ray of light $\vec{L}$, there's a single reflected ray $\vec{R}$. But not every object is perfectly polished; while *most* of the light is reflected in the direction of $\vec{R}$, *some* of it is reflected in directions close to $\vec{R}$. The closer to $\vec{R}$, the more light is reflected in that direction, as you can see in Figure 3-10. The "shininess" of the object is what determines how rapidly the reflected light decreases as you move away from $\vec{R}$.

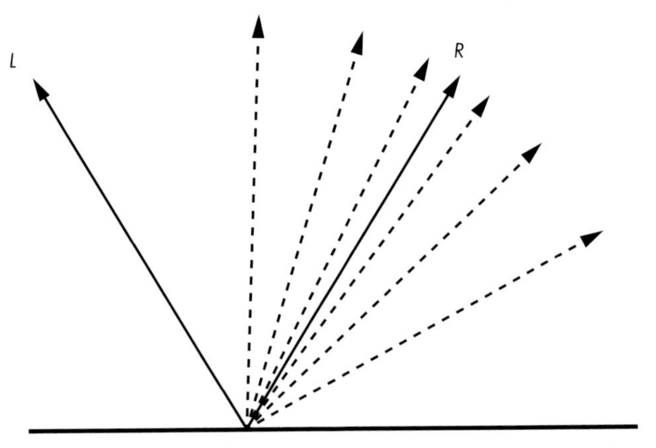

Figure 3-10: For surfaces that aren't perfectly polished, the closer a direction is to $\vec{R}$, the more rays of light are reflected in that direction.

We want to figure out how much light from $\vec{L}$ is reflected back in the direction of our point of view. If $\vec{V}$ is the "view vector" pointing from P to the camera, and α is the angle between $\vec{R}$ and $\vec{V}$, we get Figure 3-11.

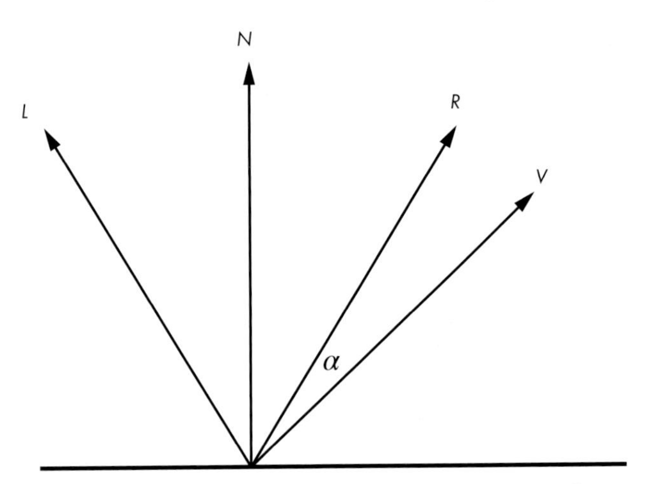

Figure 3-11: The vectors and angles involved in the specular reflection calculation

For $\alpha = 0°$, all the light is reflected in the direction of $\vec{V}$. For $\alpha = 90°$, no light is reflected. As with diffuse reflection, we need a mathematical expression to determine what happens for intermediate values of α.

Modeling Specular Reflection

At the beginning of this chapter, I mentioned that some models aren't based on physical models. This is one of them. The following model is arbitrary, but it's used because it's easy to compute and it looks good.

Consider $\cos(\alpha)$. It has the nice properties that $\cos(0) = 1$ and $\cos(\pm 90) = 0$, just like we need; and the values become gradually smaller from 0 to 90 in a very pleasant curve (Figure 3-12).

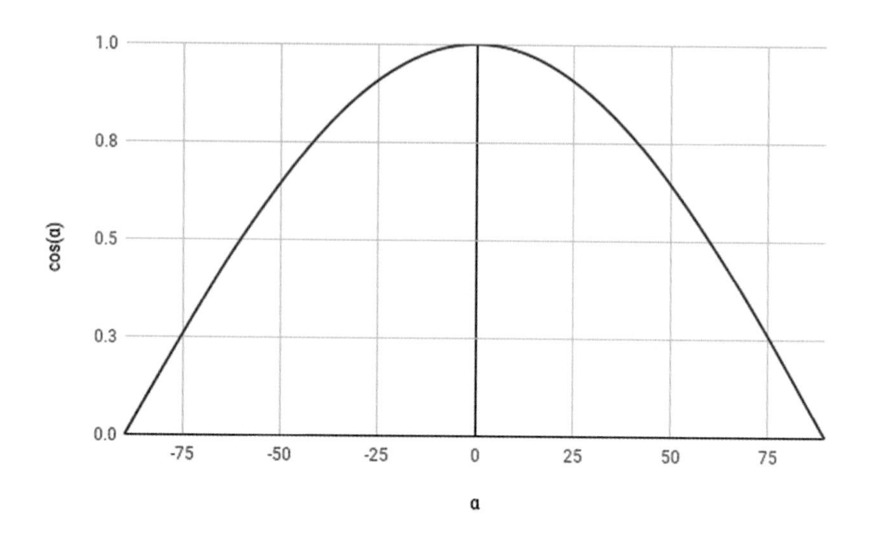

Figure 3-12: The graph of $\cos(\alpha)$.

This means $\cos(\alpha)$ matches all of our requirements for the specular reflection function, so why not use it?

There's one more detail. If we used this formula straight away, every object would be equally shiny. How can we adapt the equation to represent varying degrees of shininess?

Remember that shininess is a measure of how quickly the reflection function decreases as α increases. A simple way to obtain different shininess curves is to compute the power of $\cos(\alpha)$ to some positive exponent s. Since $0 \leq \cos(\alpha) \leq 1$, we are guaranteed that $0 \leq \cos(\alpha)^s \leq 1$; so $\cos(\alpha)^s$ is just like $\cos(\alpha)$, only "narrower." Figure 3-13 shows the graph for $\cos(\alpha)^s$ for different values of s.

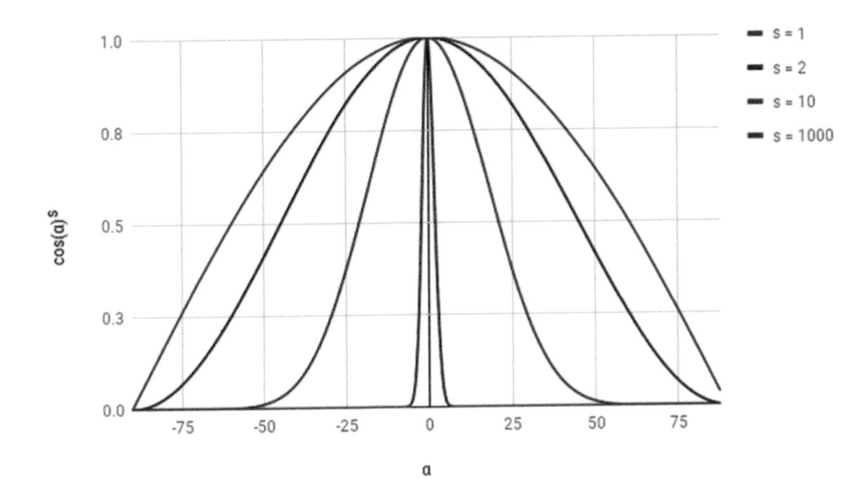

Figure 3-13: The graph of $\cos(\alpha)^s$

The bigger the value of s, the "narrower" the function becomes around 0 and the shinier the object looks. s is called the *specular exponent* and it's a property of the surface. Since the model is not based on physical reality, the values of s can only be determined by trial and error—essentially, tweaking the values until they look "right." For a physically based model, you can look into bi-directional reflectance functions (BDRFs).

Let's put all of this together. A ray of light hits a surface with specular exponent s at point P, where its normal is $\vec{N}$, from direction $\vec{L}$. How much light is reflected toward the viewing direction $\vec{V}$?

According to our model, this value is $\cos(\alpha)^s$, where α is the angle between $\vec{V}$ and $\vec{R}$; $\vec{R}$ is in turn $\vec{L}$ reflected with respect to $\vec{N}$. So the first step is to compute $\vec{R}$ from $\vec{N}$ and $\vec{L}$.

We can decompose $\vec{L}$ into two vectors, $\vec{L_P}$ and $\vec{L_N}$, such that $\vec{L} = \vec{L_N} + \vec{L_P}$, where $\vec{L_N}$ is parallel to $\vec{N}$ and $\vec{L_P}$ is perpendicular to $\vec{N}$ (Figure 3-14).

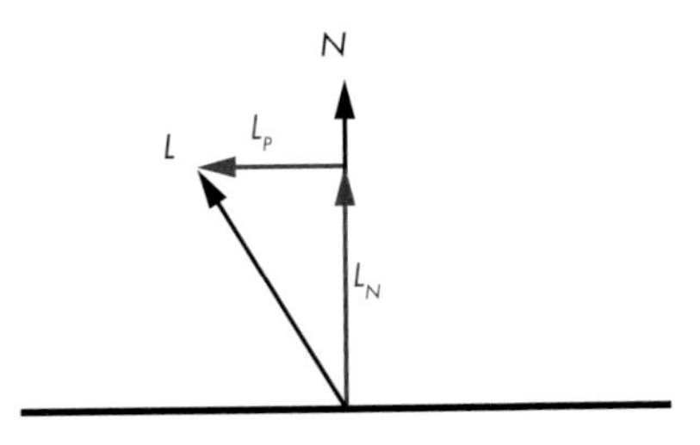

Figure 3-14: Decomposing $\vec{L}$ *into its components* $\vec{L_P}$ *and* $\vec{L_N}$

$\vec{L_N}$ is the projection of $\vec{L}$ over $\vec{N}$; by the properties of the dot product and the fact that $|\vec{N}| = 1$, the length of this projection is $\langle \vec{N}, \vec{L} \rangle$. We defined $\vec{L_N}$ to be parallel to $\vec{N}$, so $\vec{L_N} = \vec{N}\langle \vec{N}, \vec{L} \rangle$.

Since $\vec{L} = \vec{L_P} + \vec{L_N}$, we can immediately get $\vec{L_P} = \vec{L} - \vec{L_N} = \vec{L} - \vec{N}\langle \vec{N}, \vec{L} \rangle$.

Now let's look at $\vec{R}$. Since it's symmetrical to $\vec{L}$ with respect to $\vec{N}$, its component parallel to $\vec{N}$ is the same as $\vec{L}$'s, and its perpendicular component is the opposite of $\vec{L}$'s; that is, $\vec{R} = \vec{L_N} - \vec{L_P}$. You can see this in Figure 3-15.

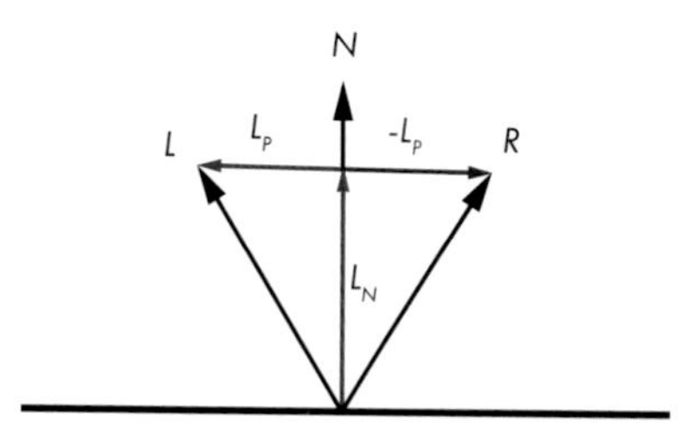

Figure 3-15: Computing $\vec{L_R}$

Substituting with the expressions we found above, we get

$$\vec{R} = \vec{N}\langle \vec{N}, \vec{L} \rangle - \vec{L} + \vec{N}\langle \vec{N}, \vec{L} \rangle$$

and simplifying a bit

$$\vec{R} = 2\vec{N}\langle \vec{N}, \vec{L} \rangle - \vec{L}$$

The Specular Reflection Term

We're now ready to write an equation for the specular reflection:

$$\vec{R} = 2\vec{N}\langle \vec{N}, \vec{L} \rangle - \vec{L}$$

$$I_S = I_L \left(\frac{\langle \vec{R}, \vec{V} \rangle}{|\vec{R}||\vec{V}|} \right)^s$$

As with diffuse lighting, it's possible that $\cos(\alpha)$ is negative, and we should ignore it for the same reason as before. Also, not every object has to be shiny; for matte objects, the specular term shouldn't be computed at all. We'll note this in the scene by setting their specular exponent to -1 and handling them accordingly.

The Full Illumination Equation

We can add the specular reflection term to the illumination equation we've been developing and get a single expression that describes illumination at a point:

$$I_P = I_A + \sum_{i=1}^{n} I_i \cdot \left[\frac{\langle \vec{N}, \vec{L_i} \rangle}{|\vec{N}||\vec{L_i}|} + \left(\frac{\langle \vec{R_i}, \vec{V} \rangle}{|\vec{R_i}||\vec{V}|} \right)^s \right]$$

where I_P is the total illumination at point P, I_A is the intensity of the ambient light, N is the normal of the surface at P, V is the vector from P to the camera, s is the specular exponent of the surface, I_i is the intensity of light i, L_i is the vector from P to light i, and R_i is the reflection vector at P for light i.

Rendering with Specular Reflections

Let's add specular reflections to the scene we've been working with so far. First, some changes to the scene itself:

```
sphere {
    center = (0, -1, 3)
    radius = 1
    color = (255, 0, 0)  # Red
    specular = 500  # Shiny
}
sphere {
    center = (2, 0, 4)
    radius = 1
    color = (0, 0, 255)  # Blue
    specular = 500  # Shiny
}
sphere {
    center = (-2, 0, 4)
    radius = 1
    color = (0, 255, 0)  # Green
    specular = 10  # Somewhat shiny
}
sphere {
    center = (0, -5001, 0)
    radius = 5000
    color = (255, 255, 0)  # Yellow
    specular = 1000  # Very shiny
}
```

This is the same scene as before, with the addition of specular exponents to the sphere definitions.

At the code level, we need to change `ComputeLighting` to compute the specular term when necessary, and add it to the overall light. Note that the function now needs $\vec{V}$ and s, as you can see in Listing 3-2.

```
ComputeLighting(P, N, V, s) {
    i = 0.0
    for light in scene.Lights {
        if light.type == ambient {
            i += light.intensity
        } else {
            if light.type == point {
                L = light.position - P
            } else {
                L = light.direction
            }

            // Diffuse
            n_dot_l = dot(N, L)
            if n_dot_l > 0 {
                i += light.intensity * n_dot_l/(length(N) * length(L))
            }

            // Specular
          ❶ if s != -1 {
                R = 2 * N * dot(N, L) - L
                r_dot_v = dot(R, V)
              ❷ if r_dot_v > 0 {
                    i += light.intensity * pow(r_dot_v/(length(R) * length(V)), s)
                }
            }
        }
    }
    return i
}
```

Listing 3-2: `ComputeLighting` that supports both diffuse and specular reflections

Most of the code remains unchanged, but we add a fragment to handle specular reflections. We make sure it applies only to shiny objects ❶ and also make sure we don't add negative light intensity ❷, as we did for diffuse reflection.

Finally, we need to modify `TraceRay` to pass the new parameters to `ComputeLighting`. s is straightforward: it comes directly from the scene definition. But where does $\vec{V}$ come from?

$\vec{V}$ is a vector that points from the object to the camera. Fortunately, we already have a vector that points from the camera to the object at `TraceRay`—that's $\vec{D}$, the direction of the ray we're tracing! So $\vec{V}$ is simply $-\vec{D}$.

Listing 3-3 gives the new `TraceRay` with specular reflection.

```
TraceRay(O, D, t_min, t_max) {
    closest_t = inf
    closest_sphere = NULL
    for sphere in scene.Spheres {
        t1, t2 = IntersectRaySphere(O, D, sphere)
        if t1 in [t_min, t_max] and t1 < closest_t {
            closest_t = t1
            closest_sphere = sphere
        }
        if t2 in [t_min, t_max] and t2 < closest_t {
            closest_t = t2
            closest_sphere = sphere
        }
    }
    if closest_sphere == NULL {
        return BACKGROUND_COLOR
    }

    P = O + closest_t * D  // Compute intersection
    N = P - closest_sphere.center  // Compute sphere normal at intersection
    N = N / length(N)
❶  return closest_sphere.color * ComputeLighting(P, N, -D, closest_sphere.specular)
}
```

Listing 3-3: TraceRay with specular reflection

The color calculation ❶ is slightly more involved than it looks. Remember that colors must be multiplied channel-wise and the results must be clamped to the range of the channel (in our case, [0–255]). Although in the example scene the light intensities add up to 1.0, now that we're adding the contributions of specular reflections, the values could go beyond that range.

You can see the reward for all this vector juggling in Figure 3-16.

Figure 3-16: The scene rendered with ambient, diffuse, and specular reflection. Not only do we get a sense of depth and volume, but each surface also has a slightly different appearance.

You can find a live implementation of this algorithm at *https://gabrielgambetta.com/cgfs/specular-demo*.

Note that in Figure 3-16, the red sphere with a specular exponent of 500 has a more concentrated bright spot than the green sphere with a specular exponent of 10, exactly as expected. The blue sphere also has a specular exponent of 500 but no visible bright spot. This is only a consequence of how the image is cropped and how the lights are placed in the scene; indeed, the left half of the red sphere also doesn't exhibit any specular reflection.

Summary

In this chapter, we've taken the very simple raytracer developed in the previous chapter and given it the ability to model lights and the way they interact with the objects in the scene.

We split lights into three types: point, directional, and ambient. We explored how each of them can represent a different type of light that you can find in real life, and how to describe them in our scene definition.

We then turned our attention to the surface of the objects in the scene, splitting them into two types: matte and shiny. We discussed how rays of light interact with them and developed two models—diffuse and specular reflection—to compute how much light they reflect toward the camera.

The end result is a much more realistic rendering of the scene: instead of seeing just the outlines of the objects, we now get a real sense of depth and volume and a feel for the materials the objects are made of.

However, we are missing a fundamental aspect of lights: shadows. This is the focus of the next chapter.

4

SHADOWS AND REFLECTIONS

Our quest to render the scene in progressively more realistic ways continues. In the previous chapter, we modeled the way rays of light interact with surfaces. In this chapter, we'll model two aspects of the way light interacts with the scene: objects casting shadows and objects reflecting on other objects.

Shadows

Where there are lights and objects, there are shadows. We have lights and objects. So where are our shadows?

Understanding Shadows

Let's begin with a more fundamental question. Why *should* there be shadows? Shadows happen when there's a light whose rays can't reach an object because there's some other object in the way.

In the previous chapter, we only looked at the very local interactions between a light source and a surface, while ignoring everything else happening in the scene. For shadows to happen, we need to take a more global view and consider the interaction between a light source, a surface we want to draw, and other objects present in the scene.

Conceptually, what we're trying to do is relatively simple. We want to add a little bit of logic that says "if there's an object between the point and the light, don't add the illumination coming from this light."

The two cases we want to distinguish are shown in Figure 4-1.

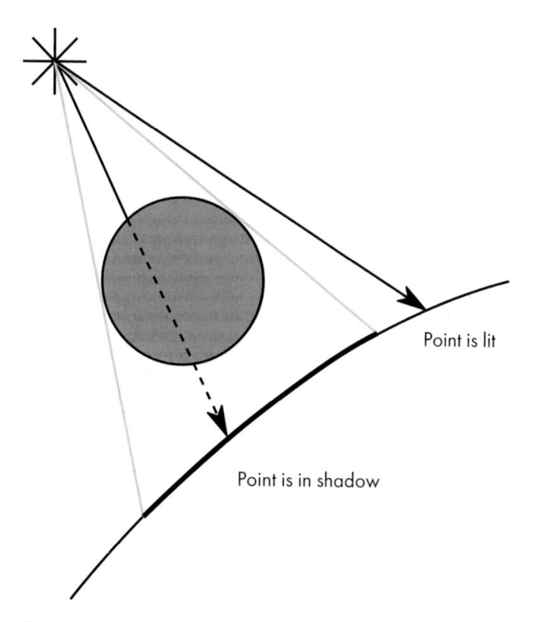

Figure 4-1: A shadow is cast over a point whenever there's an object between the light source and that point.

It turns out we already have all of the tools we need to do this. Let's start with a directional light. We know P; that's the point we're interested in. We know $\vec{L}$; that's part of the definition of the light. Knowing P and $\vec{L}$, we can define a ray, namely $P + t\vec{L}$, that goes from the point on the surface to the infinitely distant light source. Does this ray intersect any other object? If it doesn't, there's nothing between the point and the light, so we compute the illumination from this light as before. If it does, the point is in shadow, so we ignore the illumination from this light.

We already know how to compute the closest intersection between a ray and a sphere: the `TraceRay` function we're using to trace the rays from the camera. We can reuse most of it to compute the closest intersection between the ray of light and the rest of the scene.

The parameters for this function are slightly different, though:

- Instead of starting from the camera, the ray starts from P.

- The direction of the ray is not $(V - O)$ but $\vec{L}$.
- We don't want objects *behind* P to cast shadows over it, so we need $t_{min} = 0$.
- Since we're dealing with directional lights, which are infinitely far away, a very distant object should still cast a shadow over P, so $t_{max} = +\infty$.

Figure 4-2 shows two points, P_0 and P_1. When tracing a ray from P_0 in the direction of the light, we find no intersections with any objects; this means the light can reach P_0, so there's no shadow over it. In the case of P_1, we find two intersections between the ray and the sphere, with $t > 0$ (meaning the intersection is between the surface and the light); therefore, the point is in shadow.

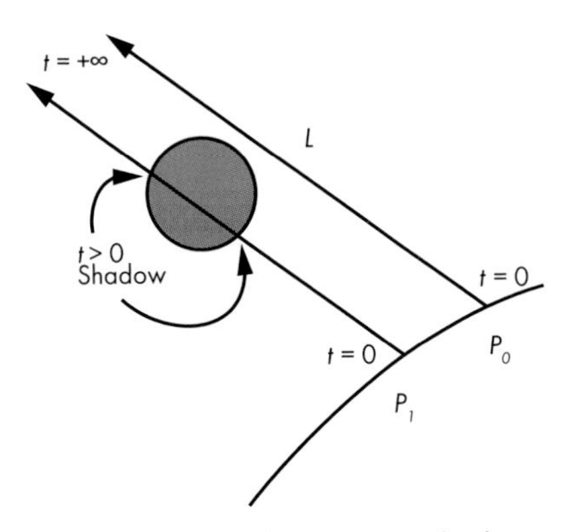

Figure 4-2: The sphere casts a shadow over P_1*, but not over* P_0*.*

We can treat point lights in a very similar way, with two exceptions. First, $\vec{L}$ is not constant, but we already know how to compute it from P and the position of the light. Second, we don't want objects farther away from the light to be able to cast a shadow over P, so in this case we need $t_{max} = 1$ so that the ray "stops" at the light.

Figure 4-3 shows these situations. When we cast a ray from P_0 with direction L_0, we find intersections with the small sphere; however, these have $t > 1$, meaning they are not between the light and P_0, so we ignore them. Therefore P_0 is not in shadow. On the other hand, the ray from P_1 with direction L_1 intersects the big sphere with $0 < t < 1$, so the sphere casts a shadow over P_1.

There's a literal edge case we need to consider. Consider the ray $P + t\vec{L}$. If we look for intersections starting from $t_{min} = 0$, we'll find one at P itself! We know P is on a sphere, so for $t = 0$, $P + 0\vec{L} = P$; in other words, every point would be casting a shadow over itself!

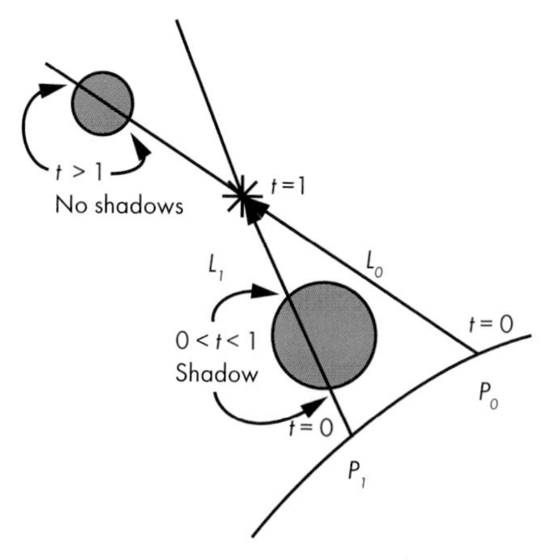

Figure 4-3: We use the value of t *at the intersections to determine whether they cast a shadow over the point.*

The simplest workaround is to set t_{min} to a very small value ϵ instead of 0. Geometrically, we're saying we want the ray to start just a tiny bit off the surface where P is, rather than exactly at P. So the range will be $[\epsilon, +\infty]$ for directional lights and $[\epsilon, 1]$ for point lights.

It might be tempting to fix this by just not computing intersections between the ray and the sphere P belongs to. This would work for spheres, but it would fail for objects with more complex shapes. For example, when you use your hand to protect your eyes from the Sun, your hand is casting a shadow over your face, and both surfaces are part of the same object - your body.

Rendering with Shadows

Let's turn the above discussion into pseudocode.

In its previous version, `TraceRay` computes the closest ray-sphere intersection, and then computes lighting on the intersection. We need to extract the closest intersection code, since we want to reuse it to compute shadows (Listing 4-1).

```
ClosestIntersection(O, D, t_min, t_max) {
    closest_t = inf
    closest_sphere = NULL
    for sphere in scene.Spheres {
        t1, t2 = IntersectRaySphere(O, D, sphere)
        if t1 in [t_min, t_max] and t1 < closest_t {
            closest_t = t1
            closest_sphere = sphere
        }
        if t2 in [t_min, t_max] and t2 < closest_t {
            closest_t = t2
```

```
                closest_sphere = sphere
            }
        }
        return closest_sphere, closest_t
    }
```

Listing 4-1: Computing the closest intersection

We can rewrite TraceRay to reuse that function, and the resulting version is much simpler (Listing 4-2).

```
TraceRay(O, D, t_min, t_max) {
    closest_sphere, closest_t = ClosestIntersection(O, D, t_min, t_max)
    if closest_sphere == NULL {
        return BACKGROUND_COLOR
    }
    P = O + closest_t * D
    N = P - closest_sphere.center
    N = N / length(N)
    return closest_sphere.color * ComputeLighting(P, N, -D, closest_sphere.specular)
}
```

Listing 4-2: A simpler version of TraceRay after factoring out ClosestIntersection

Then, we need to add the shadow check ❶ to ComputeLighting (Listing 4-3).

```
ComputeLighting(P, N, V, s) {
    i = 0.0
    for light in scene.Lights {
        if light.type == ambient {
            i += light.intensity
        } else {
            if light.type == point {
                L = light.position - P
                t_max = 1
            } else {
                L = light.direction
                t_max = inf
            }

            // Shadow check
          ❶ shadow_sphere, shadow_t = ClosestIntersection(P, L, 0.001, t_max)
            if shadow_sphere != NULL {
                continue
            }
```

```
            // Diffuse
            n_dot_l = dot(N, L)
            if n_dot_l > 0 {
                i += light.intensity * n_dot_l / (length(N) * length(L))
            }

            // Specular
            if s != -1 {
                R = 2 * N * dot(N, L) - L
                r_dot_v = dot(R, V)
                if r_dot_v > 0 {
                    i += light.intensity * pow(r_dot_v / (length(R) * length(V)), s)
                }
            }
        }
    }
    return i
}
```

Listing 4-3: `ComputeLighting` with shadow support

Figure 4-4 shows what the freshly rendered scene looks like.

Figure 4-4: A raytraced scene, now with shadows

You can find a live implementation of this algorithm at *https://gabrielgambetta.com/cgfs/shadows-demo*. In this demo, you can choose whether to trace rays from $t = 0$ or from $t = \epsilon$ to get a clearer picture of the difference it makes.

Now we're getting somewhere. Objects in the scene interact with each other in a more realistic way, casting shadows over each other. Next we'll explore more interactions between objects—namely, objects reflecting other objects.

Reflections

In the previous chapter, we talked about surfaces that are "mirror-like," but that only gave them a shiny appearance. Can we have objects that look like true mirrors—that is, can we see other objects reflected on their surface? We can, and in fact doing this in a raytracer is remarkably simple, but it can also be mind-twisting the first time you see how it's done.

Mirrors and Reflection

Let's look at how mirrors work. When you look at a mirror, what you're seeing are the rays of light that bounce off the mirror. Rays of light are reflected symmetrically with respect to the surface normal, as you can see in Figure 4-5.

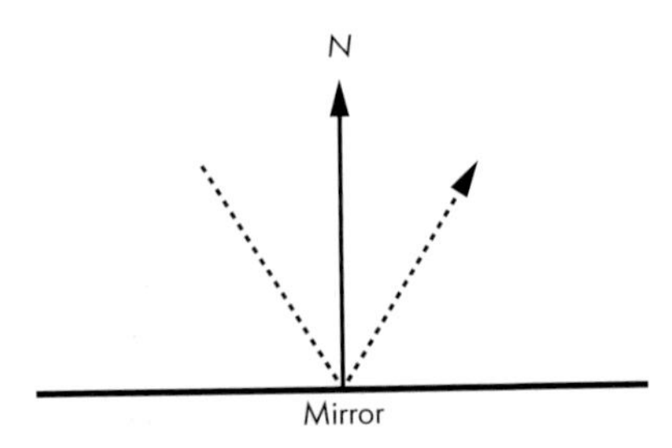

Figure 4-5: A ray of light bounces off a mirror in a direction symmetrical to the mirror's normal.

Suppose we're tracing a ray, and the closest intersection happens to be with a mirror. What color is this ray of light? It's not the color of the mirror itself, because we're looking at reflected light. So we need to figure out where this light is coming from and what color it is. So all we have to do is compute the direction of the reflected ray and figure out the color of the light coming from that direction.

If only we had a function that, given a ray, returned the color of the light coming from its direction...

Oh, wait! We do have one, and it's called `TraceRay`!

In the main loop, for each pixel, we create a ray from the camera to the scene and we call `TraceRay` to figure out what color the camera "sees" in that direction. If `TraceRay` determines that the camera is seeing a mirror, it just needs to compute the direction of the reflected ray and to figure out the color of the light coming from that direction; it must call . . . *itself*.

At this point, I suggest you read the last couple of paragraphs again until you get it. If this is the first time you've read about recursive raytracing, it may take a couple of reads and some head scratching until you really get it.

Go on, I'll wait—and once the euphoria of this beautiful *aha!* moment has started to wane, let's formalize this a bit.

When we design a recursive algorithm (one that calls itself), we need to ensure we don't cause an infinite loop (also known as "This program has stopped responding. Do you want to terminate it?"). This algorithm has two natural exit conditions: when the ray hits a non-reflective object and when it doesn't hit anything. But there's a simple case where we could get trapped in an infinite loop: the *infinite hall* effect. This is what happens when you put a mirror in front of another and look into it—infinite copies of yourself!

There are many ways to prevent an infinite recursion. We'll just introduce a *recursion limit* to the algorithm; this will control how "deep" it can go. Let's call it r. When $r = 0$, we see objects but no reflections. When $r = 1$, we see objects and the reflections of some objects on them (Figure 4-6).

Figure 4-6: Reflections limited to one recursive call (r = 1). We see spheres reflected on spheres, but the reflected spheres don't look reflective themselves.

When $r = 2$, we see objects, the reflections of some objects, and the reflections of the reflections of some objects (and so on for greater values of r). Figure 4-7 shows the result of $r = 3$. In general, it doesn't make much sense to go deeper than three levels, since the differences are barely noticeable at that point.

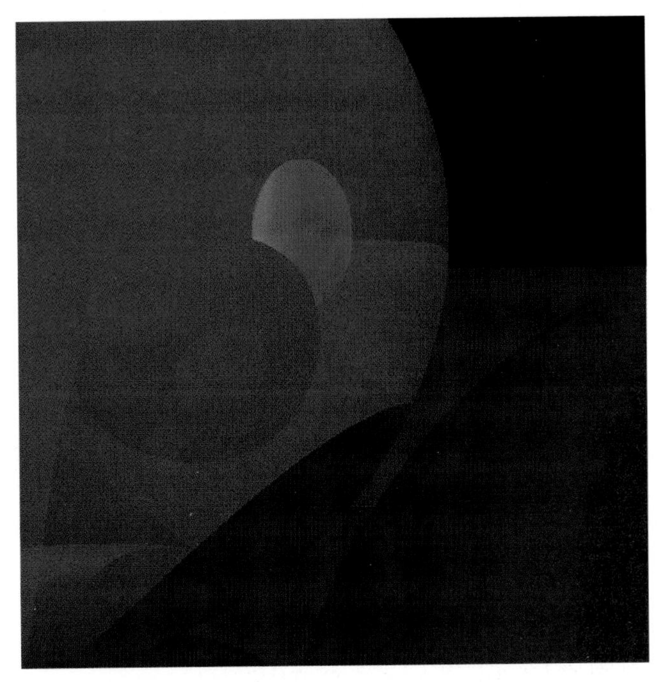

Figure 4-7: Reflections limited to three recursive calls (r = 3). Now we can see the reflections of the reflections of the reflections of the spheres.

We'll make another distinction. "Reflectiveness" doesn't have to be an all-or-nothing proposition; objects may be only partially reflective. We'll assign a number between 0 and 1 to every surface, specifying how reflective it is. Then we'll compute the weighted average of the locally illuminated color and the reflected color using that number as the weight.

Finally, what are the parameters for the recursive call to `TraceRay`?

- The ray starts at the surface of the object, P.
- The direction of the reflected ray is the direction of the incoming ray bouncing off P; in `TraceRay` we have $\vec{D}$, the direction of the incoming ray towards P, so the direction of the reflected ray is $-\vec{D}$ reflected with respect to $\vec{N}$.
- Similar to what happened with the shadows, we don't want objects to reflect themselves, so $t_{min} = \epsilon$.

- We want to see objects reflected, no matter how far away they are, so $t_{max} = +\infty$.
- The recursion limit is one less than the current recursion limit (to avoid an infinite recursion).

Now we're ready to turn this into actual pseudocode.

Rendering with Reflections

Let's add reflections to our raytracer. First, we modify the scene definition by adding a reflective property to each surface, describing how reflective it is, from 0.0 (not reflective at all) to 1.0 (a perfect mirror):

```
sphere {
    center = (0, -1, 3)
    radius = 1
    color = (255, 0, 0)  # Red
    specular = 500  # Shiny
    reflective = 0.2  # A bit reflective
}
sphere {
    center = (-2, 1, 3)
    radius = 1
    color = (0, 0, 255)  # Blue
    specular = 500  # Shiny
    reflective = 0.3  # A bit more reflective
}
sphere {
    center = (2, 1, 3)
    radius = 1
    color = (0, 255, 0)  # Green
    specular = 10  # Somewhat shiny
    reflective = 0.4  # Even more reflective
}
sphere {
    color = (255, 255, 0)  # Yellow
    center = (0, -5001, 0)
    radius = 5000
    specular = 1000  # Very shiny
    reflective = 0.5  # Half reflective
}
```

We already use the "reflect ray" formula during the computation of specular reflections, so we can factor it out. It takes a ray $\vec{R}$ and a normal $\vec{N}$ and returns $\vec{R}$ reflected with respect to $\vec{N}$.

```
ReflectRay(R, N) {
    return 2 * N * dot(N, R) - R;
}
```

The only change we need to make to ComputeLighting is replacing the reflection equation with a call to this new ReflectRay.

There's a small change in the main method—we need to pass a recursion limit to the top-level TraceRay call:

```
color = TraceRay(O, D, 1, inf, recursion_depth)
```

We can set the initial value of recursion_depth to a sensible value such as 3, as discussed previously.

The only significant changes happen near the end of TraceRay, where we compute the reflections recursively. You can see the changes in Listing 4-4.

```
TraceRay(O, D, t_min, t_max, recursion_depth) {
    closest_sphere, closest_t = ClosestIntersection(O, D, t_min, t_max)

    if closest_sphere == NULL {
        return BACKGROUND_COLOR
    }

    // Compute local color
    P = O + closest_t * D
    N = P - closest_sphere.center
    N = N / length(N)
    local_color = closest_sphere.color * ComputeLighting(P, N, -D, closest_sphere.specular)

    // If we hit the recursion limit or the object is not reflective, we're done
❶ r = closest_sphere.reflective
    if recursion_depth <= 0 or r <= 0 {
        return local_color
    }

    // Compute the reflected color
    R = ReflectRay(-D, N)
❷ reflected_color = TraceRay(P, R, 0.001, inf, recursion_depth - 1)

❸ return local_color * (1 - r) + reflected_color * r
}
```

Listing 4-4: The raytracer pseudocode, now with reflections

The changes to the code are surprisingly simple. First, we check whether we need to compute reflections at all ❶. If the sphere is not reflective or we hit the recursion limit, we're done, and we can just return the sphere's own color.

The most interesting change is the recursive call ❷; `TraceRay` calls itself, with the appropriate parameters for reflection and, importantly, decrementing the recursion depth counter; this, combined with the check ❶, prevents an infinite loop.

Finally, once we have the sphere's local color and the reflected color, we blend them together ❸, using "how reflective this sphere is" as the blending weight.

I'll let the results speak for themselves. Check out Figure 4-8.

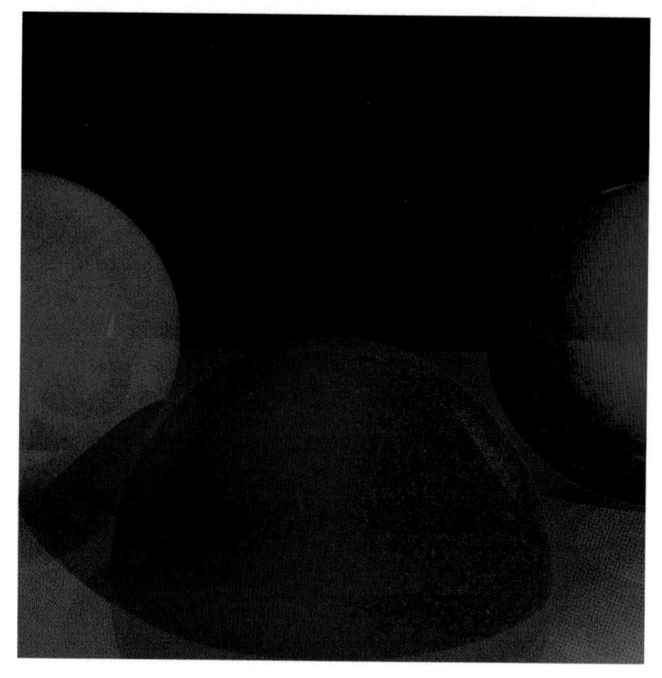

Figure 4-8: The raytraced scene, now with reflections

You can find a live implementation of this algorithm at *https://gabrielgambetta.com/cgfs/reflections-demo*.

Summary

In the previous chapters, we developed a basic framework to render a 3D scene on a 2D canvas, modeling the way a ray of light interacts with the surface of an object. This gave us a simple initial representation of the scene.

In this chapter, we extended this framework to model how different objects in the scene interact not only with rays of light, but with each other—by casting shadows over each other and by reflecting each other. As a result, the rendered scene looks significantly more realistic.

In the next chapter, we'll briefly discuss different ways to extend this work, from representing objects other than spheres to practical considerations such as rendering performance.

5

EXTENDING THE RAYTRACER

We'll conclude the first part of the book with a quick discussion of several interesting topics that we haven't yet covered: placing the camera anywhere in the scene, performance optimizations, primitives other than spheres, modeling objects using constructive solid geometry, supporting transparent surfaces, and supersampling. We won't implement all of these changes, but I encourage you to give them a try! The preceding chapters, plus the descriptions offered below, give you solid foundations to explore and implement them by yourself.

Arbitrary Camera Positioning

At the very beginning of the discussion about raytracing we made three important assumptions: that the camera was fixed at $(0, 0, 0)$, that it was pointing to $\vec{Z_+}$, and that its "up" direction was $\vec{Y_+}$. In this section, we'll lift these restrictions so we can put the camera anywhere in the scene and point it in any direction.

Let's start with the camera position. You may have noticed that O is used exactly once in all the pseudocode: as the origin of the rays coming from the camera in the top-level method. If we want to change the position of the camera, the *only* thing we need to do is to use a different value for O and we're done.

Does the change in *position* affect the *direction* of the rays? Not at all. The direction of the rays is the vector that goes from the camera to the projection plane. When we move the camera, the projection plane moves together with it, so their relative positions don't change. The way we have written `CanvasToViewport` is consistent with this idea.

Let's turn our attention to the camera orientation. Suppose you have a rotation matrix that represents the desired orientation of the camera. The *position* of the camera doesn't change if you just rotate the camera around, but the direction it's looking toward does; it undergoes the same rotation as the whole camera. So if you have the ray direction $\vec{D}$ and the rotation matrix R, the rotated D is just $R \cdot \vec{D}$.

In summary, the only function that needs to change is the main function we wrote back in Listing 2-2. Listing 5-1 shows the updated function.

```
for x in [-Cw/2, Cw/2] {
    for y in [-Ch/2, Ch/2] {
      ❶ D = camera.rotation * CanvasToViewport(x, y)
      ❷ color = TraceRay(camera.position, D, 1, inf)
        canvas.PutPixel(x, y, color)
    }
}
```

Listing 5-1: The main loop, updated to support an arbitrary camera position and orientation

We apply the camera's rotation matrix ❶, which describes its orientation in space, to the direction of the ray we're about to trace. Then we use the camera position as the starting point of the ray ❷.

Figure 5-1 shows what our scene looks like when rendered from a different position and with a different camera orientation.

Figure 5-1: Our familiar scene, rendered with a different camera position and orientation

You can find a live implementation of this algorithm at *https://gabrielgambetta.com/cgfs/camera-demo*.

Performance Optimizations

The preceding chapters focused on the clearest possible way to explain and implement the different features of a raytracer. As a result, it is fully functional but not particularly fast. Here are some ideas you can explore by yourself to make the raytracer faster. Just for fun, measure before-and-after times for each of these. You'll be surprised by the results!

Parallelization

The most obvious way to make a raytracer faster is to trace more than one ray at a time. Since each ray leaving the camera is independent of every other ray and the scene data is read-only, you can trace one ray per CPU core without many penalties or much synchronization complexity. In fact, raytracers belong to a class of algorithms called *embarrassingly parallelizable*, precisely because their very nature makes them extremely easy to parallelize.

Spawning a thread per ray is probably not a good idea, though; the overhead of managing potentially millions of threads would probably negate the speed-up you'd obtain. A more sensible idea would be to create a set of "tasks," each of them responsible for raytracing a section of the canvas (a rectangular area, down to a single pixel), and dispatch them to worker threads running on the physical cores as they become available.

Caching Immutable Values

Caching is a way to avoid repeating the same computation over and over again. Whenever there's an expensive computation and you expect to use the result of this computation repeatedly, it might be a good idea to store (cache) this result and just reuse it next time it's needed, especially if this value doesn't change often.

Consider the values computed in `IntersectRaySphere`, where a raytracer typically spends most of its time:

```
a = dot(D, D)
b = 2 * dot(OC, D)
c = dot(OC, OC) - r * r
```

Different values are immutable during different periods of time.

Once you load the scene and you know the size of the spheres, you can compute `r * r`. That value won't change unless the size of the spheres changes.

Some values are immutable for an entire frame, at the very least. One such value is `dot(OC, OC)` and it only needs to change between frames if the camera or a sphere moves. (Note that shadows and reflections trace rays that don't start at the camera, so some care is needed to make sure the cached value isn't used in that case.)

Some values don't change for an entire ray. For example, you can compute `dot(D, D)` in `ClosestIntersection` and pass it to `IntersectRaySphere`.

There are many other computations that can be reused. Use your imagination! Not every cached value will make things faster overall, however, because sometimes the bookkeeping overhead might be higher than the time saved. Always use benchmarks to evaluate whether an optimization is actually helping.

Shadow Optimizations

When a point of a surface is in shadow because there is another object in the way, it's quite likely that the point right next to it will also be in the shadow of the same object (this is called *shadow coherence*). You can see an example of this in Figure 5-2.

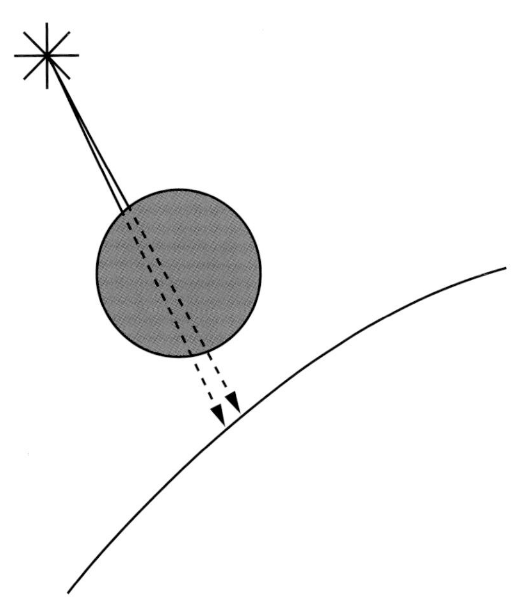

Figure 5-2: Points that are close together are likely to be in the shadow of the same object.

When searching for objects between the point and the light, to determine whether the point is in shadow, we'd normally check for intersections with every other object. However, if we know that the point immediately next to it is in the shadow of a specific object, we can check for intersections with that object first. If we find one, we're done and we don't need to check every other object! If we don't find intersections with that object, we just revert to checking every object.

In the same vein, when looking for ray-object intersections to determine whether a point is in shadow, you don't really need the *closest* intersection; it's enough to know that there's *at least one* intersection, because that will be enough to stop the light from reaching the point! So you can write a specialized version of `ClosestIntersection` that returns as soon as it finds *any* intersection. You also don't need to compute and return `closest_t`; instead, you can return just a Boolean value.

Spatial Structures

Computing the intersection of a ray with every sphere in the scene is somewhat wasteful. There are many data structures that let you discard entire groups of objects at once without having to compute the intersections individually.

Suppose you have several spheres close to each other. You can compute the center and radius of the smallest sphere that contains all these spheres. If a ray doesn't intersect this *bounding sphere*, you can be sure that it doesn't intersect any of the spheres it contains, at the cost of a single intersection test. Of course, if it does, you still need to check whether it intersects any of the spheres it contains.

You could go further and have several levels of bounding spheres (that is, groups of groups of spheres), forming a hierarchy that needs to be traversed all the way to the bottom only when there's a good chance that one of the actual spheres will be intersected by a ray.

While the exact details of this family of techniques are outside the scope of this book, you can find more information under the name *bounding volume hierarchy*.

Subsampling

Here's an easy way to make your raytracer *N* times faster: compute *N* times fewer pixels!

For each pixel in the canvas, we trace one ray through the viewport to *sample* the color of the light coming from that direction. If we had fewer rays than pixels, we'd be *subsampling* the scene. But how can we do this and still render the scene correctly?

Suppose you trace the rays for the pixels (10, 100) and (12, 100), and they happen to hit the same object. You can reasonably assume that the ray for the pixel (11, 100) will also hit the same object, so you can skip the initial search for intersections with all the objects in the scene and jump straight to computing the color at that point.

If you skip every other pixel in both the horizontal and vertical directions, you could be doing up to 75 percent fewer primary ray-scene intersection computations—that's a 4x speedup!

Of course, you may well miss a very thin object; this is an "impure" optimization, in the sense that, unlike the ones discussed before, it results in an image that closely resembles, but is not guaranteed to be identical to, the image without the optimization. In a way, it's "cheating" by cutting corners. The trick is to know what corners can be cut while maintaining satisfactory results; in many areas of computer graphics, what matters is the subjective quality of the results.

Supporting Other Primitives

In the previous chapters, we've used spheres as primitives because they're mathematically easy to manipulate; that is, the equations to find the intersections between rays and spheres are relatively simple. But once you have a basic raytracer than can render spheres, adding support to render other primitives doesn't require much additional work.

Note that `TraceRay` needs to be able to compute just two things for a ray and any given object: the value of t for the closest intersection between them and the normal at that intersection. Everything else in the raytracer is object-independent!

Triangles are a good primitive to support. A triangle is the simplest possible polygon, so you can build any other polygon out of triangles. They're mathematically easy to manipulate, so they're a good way to represent approximations of more complex surfaces.

To add triangle support to the raytracer, you only need to change `TraceRay`. First, you compute the intersection between the ray (given by its origin and direction) and the plane that contains the triangle (given by its normal and its distance from the origin).

Since planes are infinitely big, rays will almost always intersect any given plane (except if they're exactly parallel). So the second step is to determine whether the ray-plane intersection is actually inside the triangle. There are many ways to do this, including using barycentric coordinates or using cross-products to check whether the point is "on the inside" with respect to each of the three sides of the triangle.

Once you have determined that the point is inside the triangle, the normal at the intersection is just the normal of the plane. Have `TraceRay` return the appropriate values and no further changes will be required!

Constructive Solid Geometry

Suppose we want to render objects more complicated than spheres or curved objects that are difficult to model accurately using a set of triangles. Two good examples are lenses (like the ones in magnifying glasses) and the Death Star (that's no moon . . .).

We can easily describe these objects in plain language. A magnifying glass looks like two slices of a sphere glued together; the Death Star looks like a sphere with a smaller sphere taken out of it.

We can express this more formally as the result of applying set operations (such as union, intersection, or difference) to other objects. Continuing with our examples above, a lens can be described as the intersection of two spheres and the Death Star as a big sphere from which we subtract a smaller sphere (see Figure 5-3).

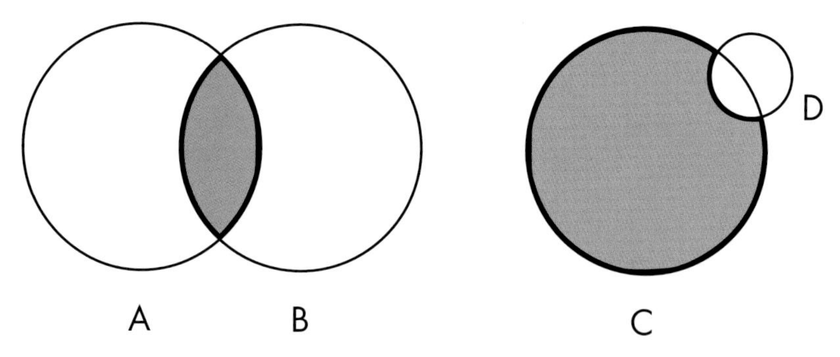

Figure 5-3: Constructive solid geometry in action. A ∩ B *gives us a lens.* C − D *gives us the Death Star.*

You might be thinking that computing Boolean operations of solid objects is a very tricky geometrical problem. And you'd be completely correct! Fortunately, it turns out that *constructive solid geometry* lets us render the results of set operations between objects without ever having to explicitly compute these results!

How can we do this in our raytracer? For every object, you can compute the points where the ray enters and exits the object; in the case of a sphere, for example, the ray enters at $min(t_1, t_2)$ and exits at $max(t_1, t_2)$. Suppose you want to compute the intersection of two spheres; the ray is inside the intersection when it's inside *both* spheres, and it's outside when it's outside *either* sphere. In the case of the subtraction, the ray is inside when it's inside the first object but not the second one. For the union of two objects, the ray is inside when it's inside either of the objects.

More generally, if you want to compute the intersection between a ray and the object $A \odot B$ (where $\odot$ is any set operation), you first compute the intersection between the ray and A and B separately, which gives you the ranges of t that are "inside" for each object, R_A and R_B. Then you compute $R_A \odot R_B$, which is the "inside" range for $A \odot B$. Once you have this, the closest intersection between the ray and $A \odot B$ is the smallest value of t that is both in the "inside" range of the object, and between t_{min} and t_{max}. Figure 5-4 shows the inside range for the union, intersection, and subtraction of two spheres.

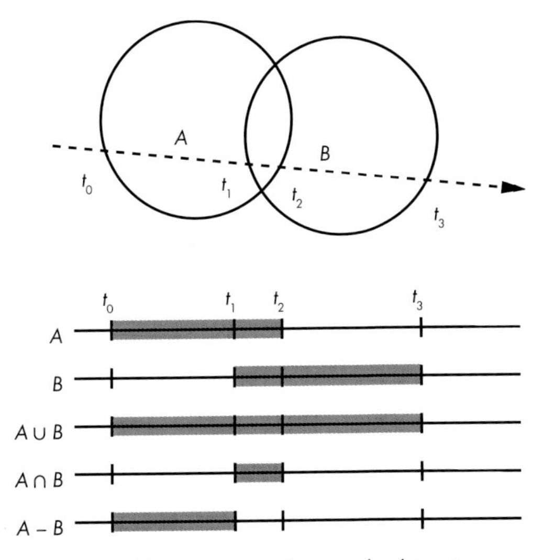

Figure 5-4: Union, intersection, and subtraction of two spheres

The normal at the intersection is either the normal of the object that produced the intersection or its opposite, depending on whether you're looking at the "outside" or "inside" of the original object.

Of course, A and B don't have to be primitives; they can be the result of set operations themselves! If you implement this cleanly, you don't even need to know *what* A and B are, as long as you can get intersections and normals out of them. This way you can take three spheres and compute, for example, $(A \cup B) \cap C$.

Transparency

So far we have rendered every object as if it were fully opaque, but this doesn't need to be the case. We can render partially transparent objects, like a fishbowl.

Implementing this is quite similar to implementing reflection. When a ray hits a partially transparent surface, you compute the local and reflected color as before, but you also compute an additional color—the color of the light coming *through* the object, obtained with another call to `TraceRay`. Then you blend this color with the local and reflected colors, depending on how transparent the object is, much in the same way we did when computing object reflections.

Refraction

In real life, when a ray of light goes through a transparent object, it changes direction (this is why when you submerge a straw in a glass of water, it looks "broken"). More precisely, a ray of light changes direction when it's going through a material (such as air) and enters a different material (such as water).

The way the direction changes depends on a property of each material, called its *refraction index*, according to the following equation, called Snell's Law:

$$\frac{\sin(\alpha_1)}{\sin(\alpha_2)} = \frac{n_2}{n_1}$$

Here, α_1 and α_2 are the angles between the ray and the normal before and after crossing the surface, and n_1 and n_2 are the refraction indices of the material outside and inside objects.

For example, n_{air} is approximately 1.0, and n_{water} is approximately 1.33. So for a ray of light entering water at a 60° angle, we have

$$\frac{\sin(60)}{\sin(\alpha_2)} = \frac{1.33}{1.0}$$

$$\sin(\alpha_2) = \frac{\sin(60)}{1.33}$$

$$\alpha_2 = \arcsin(\frac{\sin(60)}{1.33}) = 40.628°$$

This example is shown in Figure 5-5.

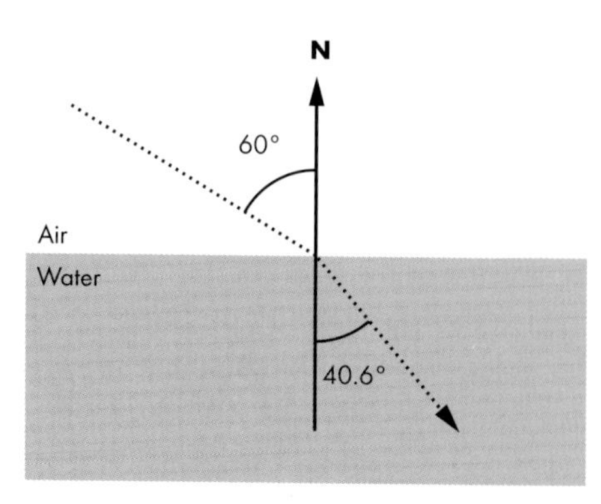

Figure 5-5: A ray of light is refracted (changes direction) as it leaves air and enters water.

At the implementation level, each ray would have to carry an additional piece of information: the refraction index of the material it is currently going through. When the ray intersects a partially transparent object, you compute the new direction of the ray from that point, based on the refraction indices of the current material and the new material, and then proceed as before.

Stop for a moment to consider this: if you implement constructive solid geometry and transparency, you can model a magnifying glass (the intersection of two spheres) that will behave like a physically correct magnifying glass!

Supersampling

Supersampling is more or less the opposite of subsampling. In this case you're looking for accuracy instead of performance. Suppose the rays corresponding to two adjacent pixels hit different objects. You would paint each pixel with the corresponding colors.

But remember the analogy that got us started: each ray is supposed to determine the "representative" color for each *square* of the "grid" we're looking through. By using a single ray per pixel, we're arbitrarily deciding that the color of the ray of light that goes through the middle of the square is representative of the whole square, but that may not be true.

The way to solve this is just to trace more rays per pixel—4, 9, 16, as many as you want—and then average them to get the color for the pixel.

Of course, this makes your raytracer 4, 9, or 16 times slower, for the exact same reasons why subsampling made it *N* times faster. Fortunately, there's a middle ground. You can assume object properties change smoothly over their surface, so shooting four rays per pixel that hit the same object at very slightly different positions may not improve the scene much. So you can start with one ray per pixel and compare adjacent rays: if they hit different objects or if the color differs by more than a certain threshold, you apply pixel subdivision to both.

Summary

In this chapter, we have briefly introduced several ideas you can explore by yourself. These modify the basic raytracer we've been developing in new and interesting ways—making it more efficient, able to represent more complex objects, or modeling rays of light in a way that better approximates our physical world.

This first part of the book should be proof that raytracers are beautiful pieces of software that can produce stunningly beautiful images using nothing but straightforward, intuitive algorithms and simple math.

Sadly, this purity comes at a cost: performance. While there are numerous way to optimize and parallelize raytracers, as discussed in this chapter, they're still too computationally expensive for real-time performance; and while hardware gets faster every year, some applications demand pictures

100 times faster—with no loss in quality. Of all these applications, games are the most demanding: we expect picture-perfect images drawn at least 60 times per second. Raytracers just don't cut it.

How have videogames been doing it since the early 90s, then?

The answer lies in a completely different family of algorithms that we'll explore in the second part of this book.

PART II

RASTERIZATION

6

LINES

In Part I of this book, we studied raytracing extensively and developed a raytracer that could render our test scene with accurate lighting, material properties, shadows, and reflection using relatively simple algorithms and math. This simplicity comes at a cost: performance. While non-real-time performance is fine for certain applications, such as architectural visualization or visual effects for movies, it's not enough for other applications, such as video games.

In this part of the book, we'll explore an entirely different set of algorithms that favor performance over mathematical purity.

Our raytracer starts from the camera and explores the scene through the viewport. For every pixel of the canvas, we answer the question, "*Which object of the scene is visible here?*" Now we'll follow an approach that is, in some sense, the opposite: for every object in the scene, we'll try to answer the question "*In which parts of the canvas will this object be visible?*"

It turns out we can develop algorithms that answer this new question much faster than raytracing could, as long as we're willing to make some accuracy trade-offs. Later, we'll explore how to use these fast algorithms to achieve results with a quality comparable to that of a raytracer.

We'll start from scratch again: we have a canvas of dimensions C_w and C_h, and we can set the color of individual pixels with `PutPixel()`, but nothing else. Let's explore how to draw the simplest possible element on the canvas: a line between two points.

Describing Lines

Suppose we have two canvas points, P_0 and P_1, with coordinates (x_0, y_0) and (x_1, y_1) respectively. How can we draw the straight line segment between P_0 and P_1?

Let's start by representing a line with parametric coordinates, just as we did with rays before (in fact, you can think of "rays" as lines in 3D). Any point P on the line can be obtained by starting at P_0 and moving some distance along the direction from P_0 to P_1:

$$P = P_0 + t(P_1 - P_0)$$

We can decompose this equation into two, one for each coordinate:

$$x = x_0 + t \cdot (x_1 - x_0)$$

$$y = y_0 + t \cdot (y_1 - y_0)$$

Let's take the first equation and solve for t:

$$x = x_0 + t \cdot (x_1 - x_0)$$

$$x - x_0 = t \cdot (x_1 - x_0)$$

$$\frac{x - x_0}{x_1 - x_0} = t$$

We can now plug this expression for t into the second equation:

$$y = y_0 + t \cdot (y_1 - y_0)$$

$$y = y_0 + \frac{x - x_0}{x_1 - x_0} \cdot (y_1 - y_0)$$

Rearranging it a bit:

$$y = y_0 + (x - x_0) \cdot \frac{y_1 - y_0}{x_1 - x_0}$$

Notice that $\frac{y_1 - y_0}{x_1 - x_0}$ is a constant that depends only on the endpoints of the segment; let's call it a. So we can rewrite the equation above as

$$y = y_0 + a \cdot (x - x_0)$$

What is a? According to the way we've defined it, it measures the change in the y coordinate per unit change in the x coordinate; in other words, it's a measure of the *slope* of the line.

Let's go back to the equation. Distributing the multiplication:

$$y = y_0 + ax - ax_0$$

Grouping the constants:

$$y = ax + (y_0 - ax_0)$$

Again, $(y_0 - ax_0)$ depends only on the endpoints of the segment; let's call it b. Finally we get

$$y = ax + b$$

This is the standard formulation of a linear function, which can be used to represent *almost* any line. When we solved for t, we added a division by $x_1 - x_0$ without thinking what happens if $x_1 = x_0$. We can't divide by zero, which means this formulation can't represent lines with $x_1 = x_0$—that is, vertical lines.

To work around this issue, we'll just ignore vertical lines for now and figure out how to deal with them later.

Drawing Lines

We now have a way to get the value of y for each value of x we're interested in. This gives us a pair (x, y) that satisfies the equation of the line.

We can now write a first approximation of a function that draws a line segment from P_0 to P_1. Let `x0` and `y0` be the x and y coordinates of P_0, respectively, and `x1` and `y1` those of P_1. Assuming $x_0 < x_1$, we can go from x_0 to x_1, computing the value of y for each value of x, and drawing a pixel at these coordinates:

```
DrawLine(P0, P1, color) {
    a = (y1 - y0) / (x1 - x0)
    b = y0 - a * x0
```

```
    for x = x0 to x1 {
        y = a * x + b
        canvas.PutPixel(x, y, color)
    }
}
```

Note that the division operator / is expected to perform real division, not integer division. This is despite x and y being integers in this context, as they represent coordinates of pixels on the canvas.

Also note that we consider `for` loops to include the last value of the range. In C, C++, Java, and JavaScript, among others, this would be written as `for (x = x0; x <= x1; ++x)`. We will be using this convention throughout this book.

This function is a direct, naive implementation of the equation above. It works, but can we make it faster?

We aren't calculating values of y for any arbitrary x. On the contrary, we're calculating them only at integer increments of x and we're doing so in order. Right after calculating $y(x)$, we calculate $y(x+1)$:

$$y(x) = ax + b$$

$$y(x+1) = a \cdot (x+1) + b$$

We can manipulate that second expression a bit:

$$y(x+1) = ax + a + b$$

$$y(x+1) = (ax + b) + a$$

$$y(x+1) = y(x) + a$$

This shouldn't be surprising; after all, the slope a is the measure of how much y changes when x increases by 1, which is exactly what we're doing here.

This means we can compute the next value of y just by taking the previous value of y and adding the slope; no per-pixel multiplication is needed, which makes the function faster. At the beginning there's no "previous value of y," so we start at (x_0, y_0). Then we keep adding 1 to x and a to y until we get to x_1.

Again assuming that $x_0 < x_1$, we can rewrite the function as follows:

```
DrawLine(P0, P1, color) {
    a = (y1 - y0) / (x1 - x0)
    y = y0
    for x = x0 to x1 {
        canvas.PutPixel(x, y, color)
        y = y + a
    }
}
```

So far we've been assuming that $x_0 < x_1$. There's an easy workaround to support lines where that doesn't hold: since the order in which we draw the pixels doesn't matter, if we get a right-to-left line, we can just swap `P0` and `P1` to transform it into the left-to-right version of the same line, and draw it as before:

```
DrawLine(P0, P1, color) {
    // Make sure x0 < x1
    if x0 > x1 {
        swap(P0, P1)
    }
    a = (y1 - y0) / (x1 - x0)
    y = y0
    for x = x0 to x1 {
        canvas.PutPixel(x, y, color)
        y = y + a
    }
}
```

Let's use our function to draw a couple of lines. Figure 6-1 shows the line segment $(-200, -100) - (240, 120)$, and Figure 6-2 shows a close-up of the line.

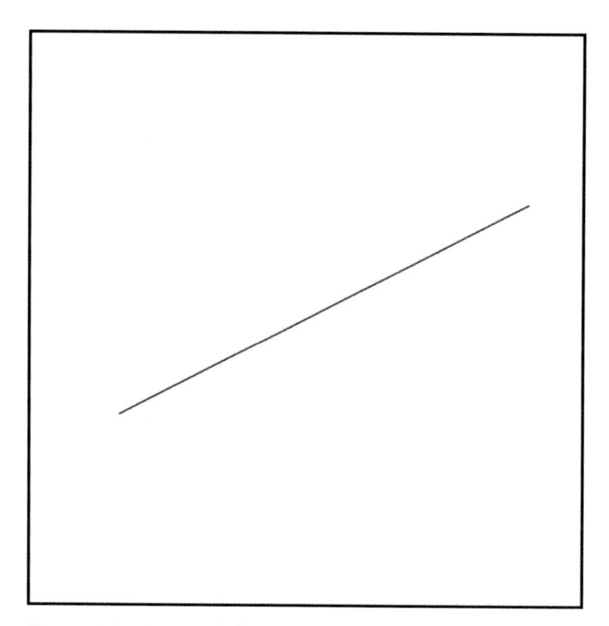

Figure 6-1: A straight line

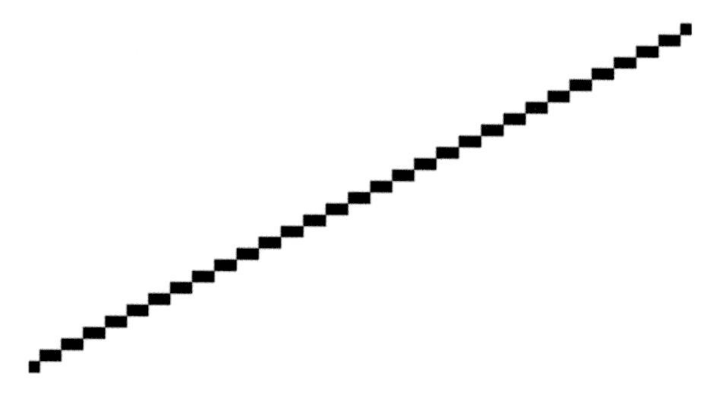

Figure 6-2: Zooming in on the straight line

The line appears jagged because we can only draw pixels on integer coordinates, and mathematical lines actually have zero width; what we're drawing is a quantized approximation of the ideal line from (−200, −100) – (240, 120). There are ways to draw prettier approximations of lines (you may want to look into MSAA, FXAA, SSAA, and TAA as possible entry points to an interesting set of rabbit holes). We won't go there for two reasons: (1) it's slower, and (2) our goal is not to draw pretty lines but to develop some basic algorithms to render 3D scenes.

Let's try another line, (−50, −200) – (60, 240). Figure 6-3 shows the result and Figure 6-4 shows the corresponding close-up.

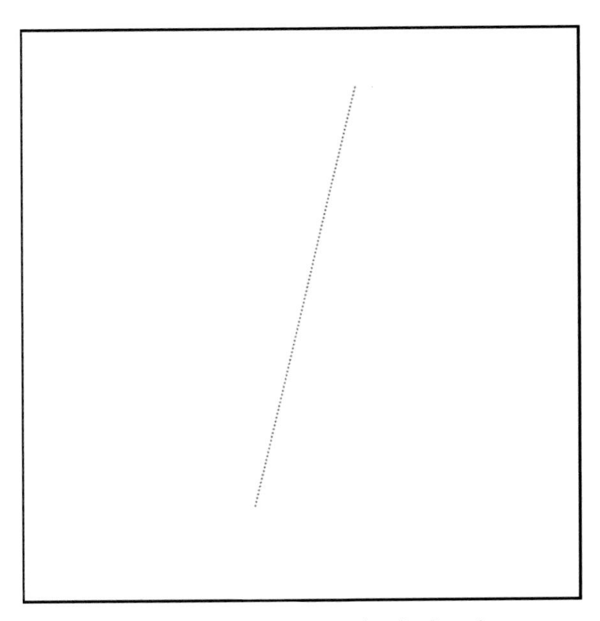

Figure 6-3: Another straight line with a higher slope

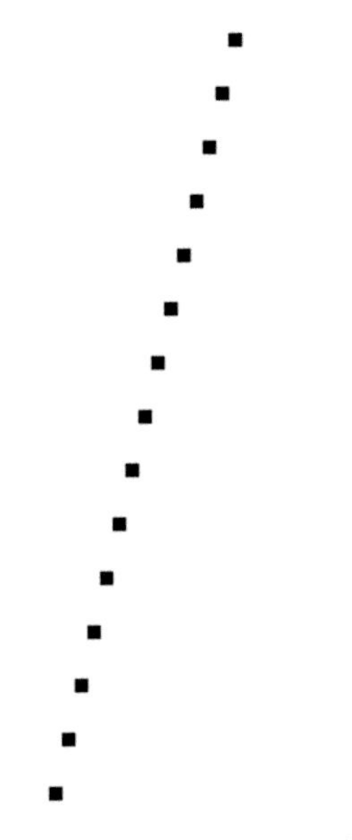

Figure 6-4: Zooming in on the second straight line

Oops. What happened?

The algorithm did exactly what we told it to; it went from left to right, computed one value of y for each value of x, and painted the corresponding pixel. The problem is that it computed *one* value of y for each value of x, while in this case we actually need *several* values of y for some values of x.

This happens because we chose a formulation where $y = f(x)$; in fact, it's the same reason why we can't draw vertical lines—an extreme case where all the values of y correspond to the same value of x.

Drawing Lines with Any Slope

Choosing $y = f(x)$ was an arbitrary choice; we could equally have chosen to express the line as $x = f(y)$. Reworking all the equations by exchanging x and y, we get the following algorithm:

```
DrawLine(P0, P1, color) {
    // Make sure y0 < y1
    if y0 > y1 {
        swap(P0, P1)
    }
    a = (x1 - x0)/(y1 - y0)
    x = x0
    for y = y0 to y1 {
        canvas.PutPixel(x, y, color)
        x = x + a
    }
}
```

This is identical to the previous `DrawLine`, except the x and y computations have been exchanged. This one can handle vertical lines and will draw $(0, 0) - (50, 100)$ correctly; but of course, it can't handle horizontal lines at all, or draw $(0, 0) - (100, 50)$ correctly! What to do?

We can just keep both versions of the function and choose which one to use depending on the line we're trying to draw. And the criterion is quite simple; does the line have more different values of x than different values of y? If there are more values of x than y, we use the first version; otherwise, we use the second.

Listing 6-1 shows a version of `DrawLine` that handles all the cases.

```
DrawLine(P0, P1, color) {
    dx = x1 - x0
    dy = y1 - y0
    if abs(dx) > abs(dy) {
        // Line is horizontal-ish
        // Make sure x0 < x1
        if x0 > x1 {
            swap(P0, P1)
        }
        a = dy/dx
```

```
        y = y0
        for x = x0 to x1 {
            canvas.PutPixel(x, y, color)
            y = y + a
        }
    } else {
        // Line is vertical-ish
        // Make sure y0 < y1
        if y0 > y1 {
            swap(P0, P1)
        }
        a = dx/dy
        x = x0
        for y = y0 to y1 {
            canvas.PutPixel(x, y, color)
            x = x + a
        }
    }
}
```

Listing 6-1: A version of `DrawLine` *that handles all the cases*

This certainly works, but it isn't pretty. There's a lot of code duplication, and the logic for selecting which function to use, the logic to compute the function values, and the pixel drawing itself are all intertwined. Surely we can do better!

The Linear Interpolation Function

We have two linear functions $y = f(x)$ and $x = f(y)$. To abstract away the fact that we're dealing with pixels, let's write it in a more generic way as $d = f(i)$, where i is the *independent variable*, the one we choose the values for, and d is the *dependent variable*, the one whose value depends on the other and which we want to compute. In the horizontal-ish case, x is the independent variable and y is the dependent variable; in the vertical-ish case, it's the other way around.

Of course, *any* function can be written as $d = f(i)$. We know two more things that completely define *our* function: the fact that it's linear and two of its values—that is, $d_0 = f(i_0)$ and $d_1 = f(i_1)$. We can write a simple function that takes these values and returns a list of all the intermediate values of d, assuming as before that $i_0 < i_1$:

```
Interpolate (i0, d0, i1, d1) {
    values = []
    a = (d1 - d0) / (i1 - i0)
```

```
    d = d0
    for i = i0 to i1 {
        values.append(d)
        d = d + a
    }
    return values
}
```

This function has the same "shape" as the first two versions of `DrawLine`, but the variables are called `i` and `d` instead of `x` and `y`, and instead of drawing pixels, this one stores the values in a list.

Note that the value of d corresponding to i_0 is returned in `values[0]`, the value for $i_0 + 1$ in `values[1]`, and so on; in general, the value for i_n is returned in `values[i_n - i_0]`, assuming i_n is in the range $[i_0, i_1]$.

There's a corner case we need to consider: we may want to compute $d = f(i)$ for a single value of i—that is, when $i_0 = i_1$. In this case we can't even compute a, so we'll treat it as a special case:

```
Interpolate (i0, d0, i1, d1) {
    if i0 == i1 {
       return [ d0 ]
    }
    values = []
    a = (d1 - d0) / (i1 - i0)
    d = d0
    for i = i0 to i1 {
        values.append(d)
        d = d + a
    }
    return values
}
```

As an implementation detail, and for the remainder of this book, the values of the independent variable i are always integers, as they represent pixels, while the values of the dependent variable d are always floating point values, as they represent values of a generic linear function.

Now we can write `DrawLine` using `Interpolate` (Listing 6-2).

```
DrawLine(P0, P1, color) {
    if abs(x1 - x0) > abs(y1 - y0) {
        // Line is horizontal-ish
        // Make sure x0 < x1
        if x0 > x1 {
            swap(P0, P1)
        }
```

```
        ys = Interpolate(x0, y0, x1, y1)
        for x = x0 to x1 {
            canvas.PutPixel(x, ys[x - x0], color)
        }
    } else {
        // Line is vertical-ish
        // Make sure y0 < y1
        if y0 > y1 {
            swap(P0, P1)
        }
        xs = Interpolate(y0, x0, y1, x1)
        for y = y0 to y1 {
            canvas.PutPixel(xs[y - y0], y, color)
        }
    }
}
```

Listing 6-2: A version of `DrawLine` *that uses* `Interpolate`

This `DrawLine` can handle all cases correctly (Figure 6-5).

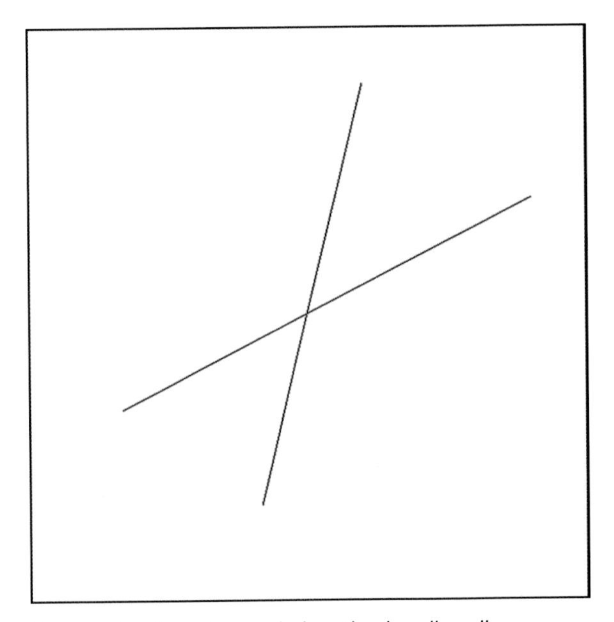

Figure 6-5: The refactored algorithm handles all cases correctly.

You can see a live demo of this refactored algorithm at *https://gabrielgambetta.com/cgfs/lines-demo.*

While this version isn't much shorter than the previous one, it cleanly separates the computation of the intermediate values of y and x from the decision of which is the independent variable and from the pixel-drawing code itself.

It might come as a surprise that this line algorithm is not the best or the fastest there is; that distinction probably belongs to *Bresenham's Algorithm*. The reason to present this algorithm is twofold. First, it is easier to understand, which is an overriding principle in this book. Second, it gave us the `Interpolate` function, which we will use extensively in the rest of this book.

Summary

In this chapter, we've taken the first steps to building a rasterizer. Using the only tool we have, `PutPixel`, we've developed an algorithm that can draw straight line segments on the canvas.

We have also developed the `Interpolate` helper method, a way to efficiently compute values of a linear function. Make sure you understand it well before proceeding, because we'll be using it a lot.

In the next chapter, we'll use `Interpolate` to draw more complex and interesting shapes on the canvas: triangles.

7

FILLED TRIANGLES

In the previous chapter, we took our first steps toward drawing simple shapes—namely, straight line segments—using only `PutPixel` and an algorithm based on simple math. In this chapter, we'll reuse some of the math to draw something more interesting: a filled triangle.

Drawing Wireframe Triangles

We can use the `DrawLine` method to draw the outline of a triangle:

```
DrawWireframeTriangle (P0, P1, P2, color) {
    DrawLine(P0, P1, color);
    DrawLine(P1, P2, color);
    DrawLine(P2, P0, color);
}
```

This kind of outline is called a *wireframe*, because it looks like a triangle made of wires, as you can see in Figure 7-1.

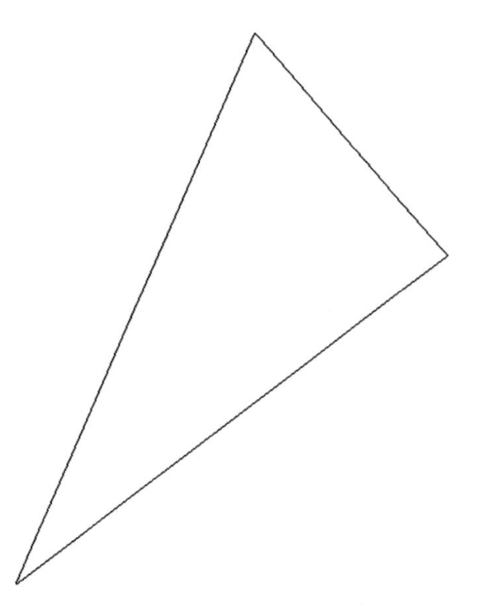

Figure 7-1: A wireframe triangle with vertices (–200,–250), (200,50), and (20,250)

This is a promising start! Next we'll explore how to fill that triangle with a color.

Drawing Filled Triangles

We want to draw a triangle filled with a color of our choice. As is often the case in computer graphics, there's more than one way to approach this problem. We'll draw filled triangles by thinking of them as a collection of horizontal line segments that look like a triangle when drawn together. Figure 7-2 shows what one such triangle would look like if we could see the individual segments.

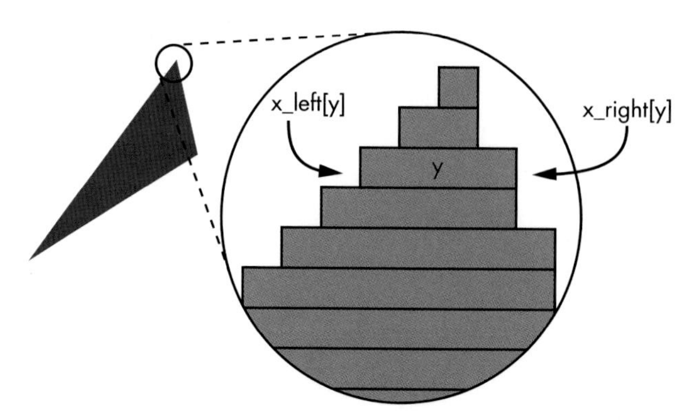

Figure 7-2: Drawing a filled triangle using horizontal segments

The following is a very rough first approximation of what we want to do:

```
for each horizontal line y between the triangle's top and bottom
    compute x_left and x_right for this y
    DrawLine(x_left, y, x_right, y)
```

Let's start with "between the triangle's top and bottom." A triangle is defined by its three vertices P_0, P_1, and P_2. If we sort these points by increasing value of y, such that $y_0 \le y_1 \le y_2$, then the range of values of y occupied by the triangle is $[y_0, y_2]$:

```
if y1 < y0 { swap(P1, P0) }
if y2 < y0 { swap(P2, P0) }
if y2 < y1 { swap(P2, P1) }
```

Sorting the vertices this way makes things easier: after doing this, we can always assume P_0 is the lowest point of the triangle and P_2 is the highest, so we won't have to deal with every possible ordering.

Next we have to compute the `x_left` and `x_right` arrays. This is slightly tricky, because the triangle has three sides, not two. However, considering only the values of y, we always have a "tall" side from P_0 to P_2, and two "short" sides from P_0 to P_1 and P_1 to P_2.

There's a special case when $y_0 = y_1$ or $y_1 = y_2$—that is, when one of the sides of the triangle is horizontal. When this happens, the two other sides have the same height, so either could be considered the "tall" side. Should we choose the right side or the left side? Fortunately, it doesn't matter; the algorithm will support both left-to-right and right-to-left horizontal lines, so we can stick to our definition that the "tall" side is the one from P_0 to P_2.

The values for `x_right` will come either from the tall side or from joining the short sides; the values for `x_left` will come from the other set. We'll start by computing the values of x for the three sides. Since we'll be drawing horizontal segments, we want exactly one value of x for each value of y; this means we can compute these values by using `Interpolate`, with y as the independent variable and x as the dependent variable:

```
x01 = Interpolate(y0, x0, y1, x1)
x12 = Interpolate(y1, x1, y2, x2)
x02 = Interpolate(y0, x0, y2, x2)
```

The x values for one of the sides are in `x02`; the values for the other side come from the concatenation of `x01` and `x12`. Note that there's a repeated value in `x01` and `x12`: the x value for y_1 is both the last value of `x01` and the first value of x12. We just need to get rid of one of them (we arbitrarily choose the last value of x01), and then concatenate the arrays:

```
remove_last(x01)
x012 = x01 + x12
```

We finally have x02 and x012, and we need to determine which is x_left and which is x_right. To do this, we can choose any horizontal line (for example, the middle one) and compare its *x* values in x02 and x012: if the *x* value in x02 is smaller than the one in x012, then we know x02 must be x_left; otherwise, it must be x_right.

```
m = floor(x02.length / 2)
if x02[m] < x012[m] {
    x_left = x02
    x_right = x012
} else {
    x_left = x012
    x_right = x02
}
```

Now we have all the data we need to draw the horizontal segments. We could use DrawLine for this. However, DrawLine is a very generic function, and in this case we're always drawing horizontal, left-to-right lines, so it's more efficient to use a simple for loop. This also gives us more "control" over every pixel we draw, which will be especially useful in the following chapters.

Listing 7-1 has the completed DrawFilledTriangle.

```
DrawFilledTriangle (P0, P1, P2, color) {
❶ // Sort the points so that y0 <= y1 <= y2
    if y1 < y0 { swap(P1, P0) }
    if y2 < y0 { swap(P2, P0) }
    if y2 < y1 { swap(P2, P1) }

❷ // Compute the x coordinates of the triangle edges
    x01 = Interpolate(y0, x0, y1, x1)
    x12 = Interpolate(y1, x1, y2, x2)
    x02 = Interpolate(y0, x0, y2, x2)
```

```
❸ // Concatenate the short sides
   remove_last(x01)
   x012 = x01 + x12

❹ // Determine which is left and which is right
   m = floor(x012.length / 2)
   if x02[m] < x012[m] {
       x_left = x02
       x_right = x012
   } else {
       x_left = x012
       x_right = x02
   }

❺ // Draw the horizontal segments
   for y = y0 to y2 {
       for x = x_left[y - y0] to x_right[y - y0] {
           canvas.PutPixel(x, y, color)
       }
   }
}
```

Listing 7-1: A function to draw filled triangles

Let's see what's going on here. The function receives the three vertices of the triangle as arguments, in any order. Our algorithm needs them to be in bottom-to-top order, so we sort them that way ❶. Next, we compute the x values for each y value of the three sides ❷, and concatenate the arrays from the two "short" sides ❸. Then we figure out which is `x_left` and which is `x_right` ❹. Finally, for each horizontal segment between the top and the bottom of the triangle, we get its left and right x coordinates, and draw the segment pixel by pixel ❺.

Figure 7-3 shows the results; for verification purposes, we call `DrawFilledTriangle` and then `DrawWireframeTriangle` with the same coordinates but different colors. Verify your results whenever you can—this is a very effective way to find bugs in the code!

Figure 7-3: A filled triangle, with wireframe edges for verification

You can find a live implementation of this algorithm at *https://gabrielgambetta.com/cgfs/triangle-demo*.

You may notice the black outline of the triangle doesn't *exactly* match the green interior region; this is especially visible in the lower-right edge of the triangle. This is because `DrawLine` is computing $y = f(x)$ for that edge but `DrawTriangle` is computing $x = f(y)$, and this can produce slightly different results due to rounding. This is the kind of approximation error we're willing to accept in order to make our rendering algorithms fast.

Summary

In this chapter, we've developed an algorithm to draw a filled triangle on the canvas. This is a step up from drawing line segments. We've also learned to think of triangles as a set of horizontal segments that we can work with individually.

In the next chapter, we'll extend the math and the algorithm to draw a triangle filled with a color gradient; the math and the reasoning behind the algorithm will be key to the rest of the features developed in this book.

8

SHADED TRIANGLES

In the previous chapter, we developed an algorithm to draw a triangle filled with a solid color. Our goal for this chapter is to draw a *shaded* triangle—that is, a triangle filled with a color gradient.

Defining Our Problem

We want to fill the triangle with different *shades* of a single color. It will look like Figure 8-1.

We need a more formal definition of what we're trying to draw. We have a base color C: for example, $0, 255, 0$, pure green. We'll assign a real value h to each vertex, denoting the intensity of the color at the vertex. h is in the $[0.0, 1.0]$ range, where 0.0 represents the darkest possible shade (that is, black) and 1.0 represents the brightest possible shade (that is, the original color—not white!).

Figure 8-1: A shaded triangle

To compute the exact color shade of a pixel given the base color of the triangle C and the intensity at that pixel h, we'll multiply channel-wise: $C_h = (R_C \cdot h, G_C \cdot h, B_C \cdot h)$. Therefore $h = 0.0$ yields pure black, $h = 1.0$ yields the original color C, and $h = 0.5$ yields a color half as bright as the original one.

Computing Edge Shading

In order to draw a shaded triangle, all we need to do is compute a value of h for each pixel of the triangle, compute the corresponding shade of the color, and paint the pixel. Easy!

At this point, however, we only know the values of h for the triangle vertices, because we chose them. How do we compute values of h for the rest of the triangle?

Let's start with the edges of the triangle. Consider the edge AB. We know h_A and h_B. What happens at M, the midpoint of AB? Since we want the intensity to vary smoothly from A to B, the value of h_M must be between h_A and h_B. Since M is in the middle of AB, why not choose h_M to be in the middle of h_A and h_B—that is, their average?

More formally, we have a function $h = f(P)$ that gives each point P an intensity value h; we know its values at A and B, $h(A) = h_A$ and $h(B) = h_B$, respectively. We want this function to be smooth. Since we know nothing else about $h = f(P)$, we can choose any function that is compatible with what we *do* know, such as a linear function (Figure 8-2).

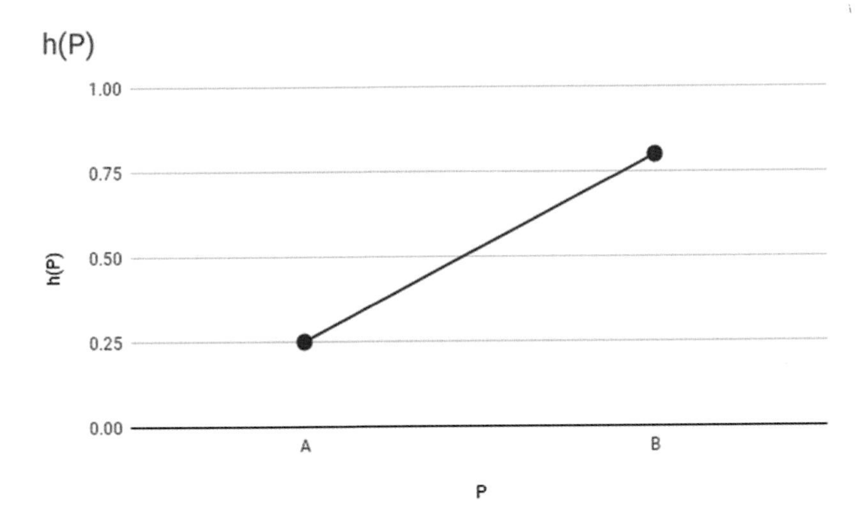

Figure 8-2: A linear function h(P)*, compatible with what we know about* h(A) *and* h(B)

This is suspiciously similar to the situation in the previous chapter: we had a linear function $x = f(y)$, we knew the values of this function at the vertices of the triangle, and we wanted to compute values of x along its sides. We can compute values of h along the sides of the triangle in a very similar way, using `Interpolate` with y as the independent variable (the values we know) and h as the dependent variable (the values we want):

```
x01 = Interpolate(y0, x0, y1, x1)
h01 = Interpolate(y0, h0, y1, h1)

x12 = Interpolate(y1, x1, y2, x2)
h12 = Interpolate(y1, h1, y2, h2)

x02 = Interpolate(y0, x0, y2, x2)
h02 = Interpolate(y0, h0, y2, h2)
```

Next, we concatenated the x arrays for the "short" sides and then determined which of `x02` and `x012` was `x_left` and which was `x_right`. Again, we can do something very similar here for the h vectors.

However, we will always use the x values to determine which side is left and which side is right, and the h values will just "follow along." x and h are properties of actual points on the screen, so we can't freely mix-and-match left- and right-side values.

We can code this as follows:

```
// Concatenate the short sides
remove_last(x01)
```

```
x012 = x01 + x12

remove_last(h01)
h012 = h01 + h12

// Determine which is left and which is right
m = floor(x012.length / 2)
if x02[m] < x012[m] {
    x_left = x02
    h_left = h02

    x_right = x012
    h_right = h012
} else {
    x_left = x012
    h_left = h012

    x_right = x02
    h_right = h02
}
```

This is very similar to the relevant section of the code in the previous chapter (Listing 7-1), except that every time we do something with an `x` vector, we do the same with the corresponding `h` vector.

Computing Interior Shading

The last step is drawing the actual horizontal segments. For each segment, we know `x_left` and `x_right`, as in the previous chapter; now we also know `h_left` and `h_right`. But this time we can't just iterate from left to right and draw every pixel with the base color: we need to compute a value of h for *each pixel* of the segment.

Again, we can assume h varies linearly with x, and use `Interpolate` to compute these values. In this case, the independent variable is x, and it goes from the `x_left` value to the `x_right` value of the specific horizontal segment we're shading; the dependent variable is h, and its corresponding values for `x_left` and `x_right` are `h_left` and `h_right` for that segment:

```
x_left_this_y = x_left[y - y0]
h_left_this_y = h_left[y - y0]

x_right_this_y = x_right[y - y0]
h_right_this_y = h_right[y - y0]
```

```
h_segment = Interpolate(x_left_this_y, h_left_this_y,
                        x_right_this_y, h_right_this_y)
```

Or, expressed in a more compact way:

```
h_segment = Interpolate(x_left[y - y0], h_left[y - y0],
                        x_right[y - y0], h_right[y - y0])
```

Now it's just a matter of computing the color for each pixel and painting it! Listing 8-1 shows the complete pseudocode for DrawShadedTriangle.

```
DrawShadedTriangle (P0, P1, P2, color) {
❶ // Sort the points so that y0 <= y1 <= y2
    if y1 < y0 { swap(P1, P0) }
    if y2 < y0 { swap(P2, P0) }
    if y2 < y1 { swap(P2, P1) }

    // Compute the x coordinates and h values of the triangle edges
    x01 = Interpolate(y0, x0, y1, x1)
    h01 = Interpolate(y0, h0, y1, h1)

    x12 = Interpolate(y1, x1, y2, x2)
    h12 = Interpolate(y1, h1, y2, h2)

    x02 = Interpolate(y0, x0, y2, x2)
    h02 = Interpolate(y0, h0, y2, h2)

    // Concatenate the short sides
    remove_last(x01)
    x012 = x01 + x12

    remove_last(h01)
    h012 = h01 + h12

    // Determine which is left and which is right
    m = floor(x012.length / 2)
    if x02[m] < x012[m] {
        x_left = x02
        h_left = h02

        x_right = x012
        h_right = h012
    } else {
        x_left = x012
        h_left = h012
```

```
        x_right = x02
        h_right = h02
    }

    // Draw the horizontal segments
 ❷ for y = y0 to y2 {
        x_l = x_left[y - y0]
        x_r = x_right[y - y0]

      ❸ h_segment = Interpolate(x_l, h_left[y - y0], x_r, h_right[y - y0])
        for x = x_l to x_r {
          ❹ shaded_color = color * h_segment[x - x_l]
            canvas.PutPixel(x, y, shaded_color)
        }
    }
}
```

Listing 8-1: A function for drawing shaded triangles

The pseudocode for this function is very similar to that for the function developed in the previous chapter (Listing 7-1). Before the horizontal segment loop ❷, we manipulate the x vectors and the h vectors in similar ways, as explained above. Inside the loop, we have an extra call to `Interpolate` ❸ to compute the h values for every pixel in the current horizontal segment. Finally, in the inner loop we use the interpolated values of h to compute a color for each pixel ❹.

Note that we're sorting the triangle vertices as before ❶. However, we now consider these vertices and their attributes, such as the intensity value h, to be an indivisible whole; that is, swapping the coordinates of two vertices must also swap their attributes.

You can find a live implementation of this algorithm at *https://gabrielgambetta.com/cgfs/gradient-demo*.

Summary

In this chapter, we've extended the triangle-drawing code developed in the previous chapter to support smoothly shaded triangles. Note that we can still use it to draw single color triangles by using 1.0 as the value of h for all three vertices.

The idea behind this algorithm is actually more general than it seems. The fact that h is an intensity value has no impact on the "shape" of the algorithm; we assign meaning to this value only at the very end, when we're about to call `PutPixel`. This means we could use this algorithm to compute the value of any *attribute* of the vertices of the triangle, for every pixel of the triangle, as long as we assume this value varies linearly on the screen.

We will indeed use this algorithm to improve the visual appearance of our triangles in the upcoming chapters. For this reason, it's a good idea to make sure you really understand this algorithm before proceeding further.

In the next chapter, however, we take a small detour. Having mastered the drawing of triangles on a 2D canvas, we will turn our attention to the third dimension.

9

PERSPECTIVE PROJECTION

So far, we have learned to draw 2D triangles on the canvas, given the 2D coordinates of their vertices. However, the goal of this book is to render 3D scenes. So in this chapter, we'll take a break from 2D triangles and focus on how to turn 3D scene coordinates into 2D canvas coordinates. We'll then use this to draw 3D triangles on the 2D canvas.

Basic Assumptions

Just like we did at the beginning of Chapter 2, we'll start by defining a *camera*. We'll use the same conventions as before: the camera is at $O = (0, 0, 0)$, looking in the direction of $\vec{Z_+}$, and its "up" vector is $\vec{Y_+}$. We'll also define a rectangular *viewport* of size V_w and V_h whose edges are parallel to $\vec{X}$ and $\vec{Y}$, at a distance d from the camera. The goal is to draw on the canvas whatever the camera sees through the viewport. If you need a refresher on these concepts, refer to Chapter 2.

Consider a point P somewhere in front of the camera. We're interested in finding P', the point on the viewport through which the camera sees P, as shown in Figure 9-1.

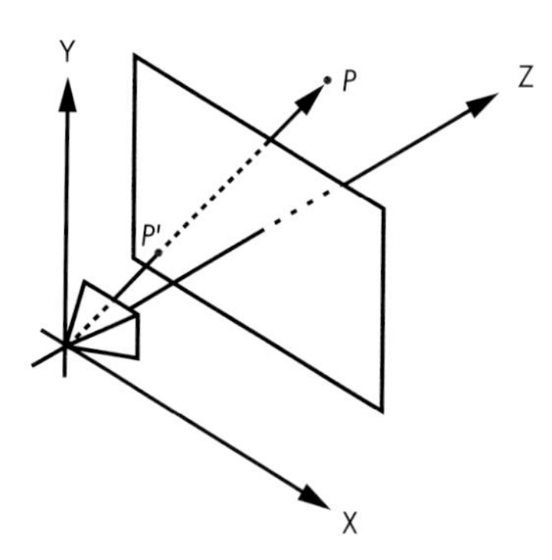

Figure 9-1: A simple perspective projection setup. The camera sees P *through* P'*, which is on the projection plane.*

This is the opposite of what we did with raytracing. Our raytracer started with a point in the canvas, and determined what it could see through that point; here, we start from a point in the scene and want to determine where it is seen on the viewport.

Finding P′

To find P', let's look at the setup shown in Figure 9-1 from a different angle, literally. Figure 9-2 shows a diagram of the setup viewed from the "right," as if we were standing on the $\vec{X}$ axis: $\vec{Y_+}$ points up, $\vec{Z_+}$ points to the right, and $\vec{X_+}$ points at us.

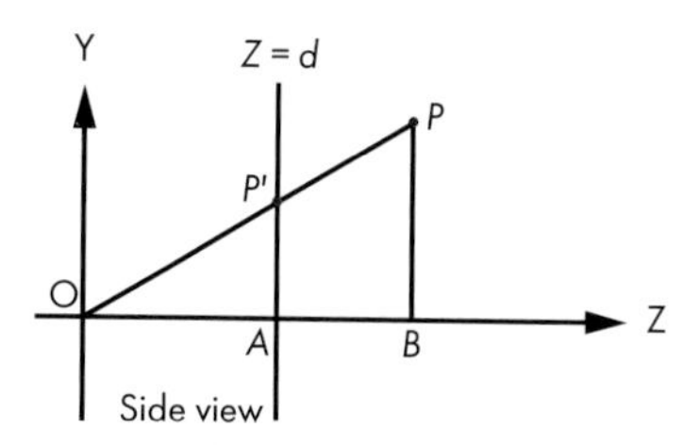

Figure 9-2: The perspective projection setup, viewed from the right

In addition to O, P, and P', this diagram also shows the points A and B, which help us reason about it.

We know that $P'_z = d$ because we defined P' to be a point on the viewport, and we know the viewport is embedded in the plane $Z = d$.

We can also see that the triangles $OP'A$ and OPB are similar, because their corresponding sides ($P'A$ and PB, OP and OP', and OA and OB) are parallel. This implies that the proportions of their sides are the same; for example:

$$\frac{|P'A|}{|OA|} = \frac{|PB|}{|OB|}$$

From that, we get

$$|P'A| = \frac{|PB| \cdot |OA|}{|OB|}$$

The (signed) length of each segment in that equation is a coordinate of a point we know or we're interested in: $|P'A| = P'_y$, $|PB| = P_y$, $|OA| = P'_z = d$, and $|OB| = P_z$. If we substitute these in the equation we get

$$P'_y = \frac{P_y \cdot d}{P_z}$$

We can draw a similar diagram, this time viewing the setup from above: $\vec{Z_+}$ points up, $\vec{X_+}$ points to the right, and $\vec{Y_+}$ points at us (Figure 9-3).

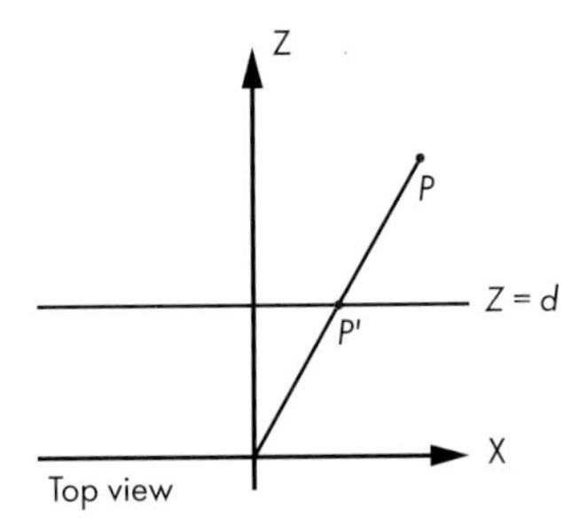

Figure 9-3: Top view of the perspective projection setup

Using similar triangles again in the same way, we can deduce that

$$P'_x = \frac{P_x \cdot d}{P_z}$$

We now have all three coordinates of P'.

The Projection Equation

Let's put all this together. Given a point P in the scene and a standard camera and viewport setup, we can compute the projection of P on the viewport, which we call P', as follows:

$$P'_x = \frac{P_x \cdot d}{P_z}$$

$$P'_y = \frac{P_y \cdot d}{P_z}$$

$$P'_z = d$$

P' is on the viewport, but it's still a point in 3D space. How do we get the corresponding point in the canvas?

We can immediately drop P'_z, because every projected point is on the viewport plane. Next we need to convert P'_x and P'_y to canvas coordinates C_x and C_y. P' is still a point in the scene, so its coordinates are expressed in scene units. We can divide them by the width and height of the viewport. These are also expressed in scene units, so we obtain temporarily unit-less values. Finally, we multiply them by the width and height of the canvas, expressed in pixels:

$$C_x = \frac{P'_x \cdot C_w}{V_w}$$

$$C_y = \frac{P'_y \cdot C_h}{V_h}$$

This viewport-to-canvas transform is the exact inverse of the canvas-to-viewport transform we used in the raytracing part of this book. And with this, we can finally go from a point in the scene to a pixel on the screen!

Properties of the Projection Equation

Before we move on, there are some interesting properties of the projection equation that are worth discussing.

The equations above should be compatible with our day-to-day experience of looking at things in the real world. For example, the farther away an object is, the smaller it looks; and indeed, if we increase P_z, we get smaller values of P'_x and P'_y.

However, things stop being so intuitive when we decrease the value of P_z too much; for negative values of P_z, that is, when an object is *behind* the camera, the object is still projected, but upside down! And, of course, when $P_z = 0$ we'd divide by zero and the universe would implode. We'll need to find a way to avoid these unpleasant situations; for now, we'll assume that every point is in front of the camera and deal with this in a later chapter.

Another fundamental property of the perspective projection is that it preserves point alignment: if three points are aligned in space, their projections will be aligned on the viewport. In other words, a straight line is always projected as a straight line. This might sound too obvious to be worth mentioning, but note, for example, that the *angle* between two lines isn't preserved: in real life, we see parallel lines "converge" at the horizon, such as when driving on a highway.

The fact that a straight line is always projected as a straight line is extremely convenient for us: so far we have talked about projecting a point, but how about projecting a line segment, or even a triangle? Because of this property, the projection of a line segment between two points is the line segment between the projection of two points; and the projection of a triangle is the triangle formed by the projections of its vertices.

Projecting Our First 3D Object

This means we can go ahead and draw our first 3D object: a cube. We define the coordinates of its 8 vertices, and we draw line segments between the projections of the 12 pairs of vertices that make the edges of the cube, as seen in Listing 9-1.

```
ViewportToCanvas(x, y) {
  return (x * Cw/Vw, y * Ch/Vh);
}

ProjectVertex(v) {
  return ViewportToCanvas(v.x * d / v.z, v.y * d / v.z)
}

// The four "front" vertices
vAf = [-1,  1, 1]
vBf = [ 1,  1, 1]
vCf = [ 1, -1, 1]
vDf = [-1, -1, 1]

// The four "back" vertices
vAb = [-1,  1, 2]
vBb = [ 1,  1, 2]
vCb = [ 1, -1, 2]
vDb = [-1, -1, 2]

// The front face
DrawLine(ProjectVertex(vAf), ProjectVertex(vBf), BLUE);
DrawLine(ProjectVertex(vBf), ProjectVertex(vCf), BLUE);
DrawLine(ProjectVertex(vCf), ProjectVertex(vDf), BLUE);
DrawLine(ProjectVertex(vDf), ProjectVertex(vAf), BLUE);

// The back face
DrawLine(ProjectVertex(vAb), ProjectVertex(vBb), RED);
```

```
DrawLine(ProjectVertex(vBb), ProjectVertex(vCb), RED);
DrawLine(ProjectVertex(vCb), ProjectVertex(vDb), RED);
DrawLine(ProjectVertex(vDb), ProjectVertex(vAb), RED);

// The front-to-back edges
DrawLine(ProjectVertex(vAf), ProjectVertex(vAb), GREEN);
DrawLine(ProjectVertex(vBf), ProjectVertex(vBb), GREEN);
DrawLine(ProjectVertex(vCf), ProjectVertex(vCb), GREEN);
DrawLine(ProjectVertex(vDf), ProjectVertex(vDb), GREEN);
```

Listing 9-1: Drawing a cube

We get something like Figure 9-4.

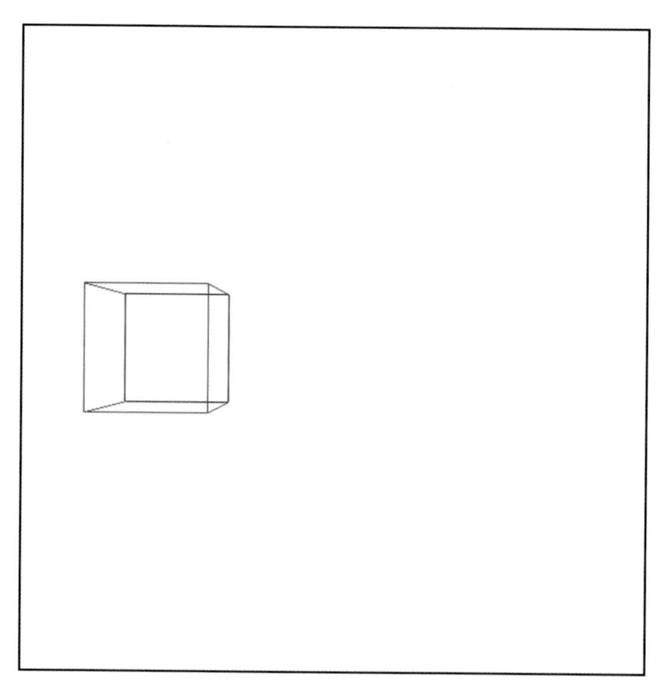

Figure 9-4: Our first 3D object projected on a 2D canvas: a cube

You can find a live implementation of this algorithm at *https://gabrielgambetta.com/cgfs/perspective-demo*.

Success! We've managed to go from the geometrical 3D representation of an object to its 2D representation as seen from our synthetic camera!

Our approach is very artisanal, though. It has many limitations. What if we want to render *two* cubes? Would we have to duplicate most of the code? What if we want to render something other than a cube? What if we want to let the user load a 3D model from a file? We clearly need a more data-driven approach to representing 3D geometry.

Summary

In this chapter, we've developed the math to go from a 3D point in the scene to a 2D point on the canvas. Because of the properties of the perspective projection, we can immediately extend this to projecting line segments and then to 3D objects.

But we have left two important issues unresolved. First, Listing 9-1 mixes the perspective projection logic with the geometry of the cube; this approach clearly won't scale. Second, because of the limitations of the perspective projection equation, it can't handle objects that are behind the camera. We will address these issues in the next two chapters.

10

DESCRIBING AND RENDERING A SCENE

In the last few chapters, we've developed algorithms to draw 2D triangles on the canvas given their 2D coordinates, and we've explored the math required to transform the 3D coordinates of points in the scene to the 2D coordinates of points on the canvas.

At the end of the previous chapter, we cobbled together a program that used both to render a 3D cube on the 2D canvas. In this chapter, we'll formalize and extend that work with the goal of rendering a whole scene containing an arbitrary number of objects.

Representing a Cube

Let's think again about how to represent and manipulate a cube, this time with the goal of finding a more general approach. The edges of our cube are 2 units long and are parallel to the coordinate axes, and it's centered on the origin, as shown in Figure 10-1.

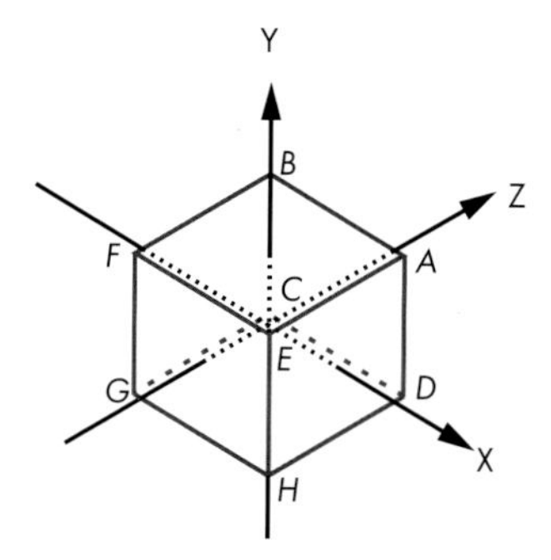

Figure 10-1: Our standard cube

These are the coordinates of its vertices:

$A = (1, 1, 1)$

$B = (-1, 1, 1)$

$C = (-1, -1, 1)$

$D = (1, -1, 1)$

$E = (1, 1, -1)$

$F = (-1, 1, -1)$

$G = (-1, -1, -1)$

$H = (1, -1, -1)$

The sides of the cube are square, but the algorithms we have developed work with triangles. One of the reasons we chose triangles in the first place is that any other polygon, including squares, can be decomposed into triangles. So we'll represent each square side of the cube using two triangles.

However, we can't take *any* three vertices of the cube and expect them to describe a triangle on its surface (for example, ADG is inside the cube). This means that the vertex coordinates, by themselves, don't fully describe the cube: we also need to know which sets of three vertices describe the triangles that make up its sides.

Here's a possible list of triangles for our cube:

```
A, B, C
A, C, D
E, A, D
E, D, H
F, E, H
F, H, G
B, F, G
B, G, C
E, F, B
E, B, A
```

```
C, G, H
C, H, D
```

This suggests a generic structure we can use to represent *any* object made of triangles: a `Vertices` list, holding the coordinates of each vertex; and a `Triangles` list, specifying which sets of three vertices describe triangles on the surface of the object.

Each entry in the `Triangles` list may include additional information besides the vertices that make it up; for example, this would be the perfect place to specify the color of each triangle.

Since the most natural way to store this information is in two lists, we'll use list indices to refer to the vertices in the vertex list. So our cube would be represented like this:

```
Vertices
0 = ( 1,  1,  1)
1 = (-1,  1,  1)
2 = (-1, -1,  1)
3 = ( 1, -1,  1)
4 = ( 1,  1, -1)
5 = (-1,  1, -1)
6 = (-1, -1, -1)
7 = ( 1, -1, -1)

Triangles
 0 = 0, 1, 2, red
 1 = 0, 2, 3, red
 2 = 4, 0, 3, green
 3 = 4, 3, 7, green
 4 = 5, 4, 7, blue
 5 = 5, 7, 6, blue
 6 = 1, 5, 6, yellow
 7 = 1, 6, 2, yellow
 8 = 4, 5, 1, purple
 9 = 4, 1, 0, purple
10 = 2, 6, 7, cyan
11 = 2, 7, 3, cyan
```

Rendering an object with this representation is quite simple: we first project every vertex, storing them in a temporary projected vertices list (since each vertex is used an average of four times, this avoids a lot of repeated work); then we go through the triangle list, rendering each individual triangle. A first approximation would look like Listing 10-1.

```
RenderObject(vertices, triangles) {
    projected = []
    for V in vertices {
        projected.append(ProjectVertex(V))
    }
    for T in triangles {
        RenderTriangle(T, projected)
    }
}

RenderTriangle(triangle, projected) {
    DrawWireframeTriangle(projected[triangle.v[0]],
                          projected[triangle.v[1]],
                          projected[triangle.v[2]],
                          triangle.color)
}
```

Listing 10-1: An algorithm to render any object made of triangles

We can go ahead and apply this directly to the cube as defined above, but the results won't look good. This is because some of its vertices are behind the camera, which, as we discussed in the previous chapter, is a recipe for weird things. And if you look at the vertex coordinates and Figure 10-1, you'll notice the coordinate origin, the position of our camera, is *inside* the cube.

To work around this problem, we'll just move the cube. To do this, we need to move each vertex of the cube in the same direction. Let's call this direction $\vec{T}$, for "translation." We'll translate the cube 7 units forward to make sure it's completely in front of the camera. We'll also translate it 1.5 units to the left to make it look more interesting. Since "forward" is the direction of $\vec{Z_+}$ and "left" is $\vec{X_-}$, the translation vector is simply

$$\vec{T} = \begin{pmatrix} -1.5 \\ 0 \\ 7 \end{pmatrix}$$

To compute the translated version V' of each vertex V in the cube, we just need to add the translation vector to it:

$$V' = V + \vec{T}$$

At this point, we can take the cube, translate each vertex, and then apply the algorithm in Listing 10-1 to get our first 3D cube (Figure 10-2).

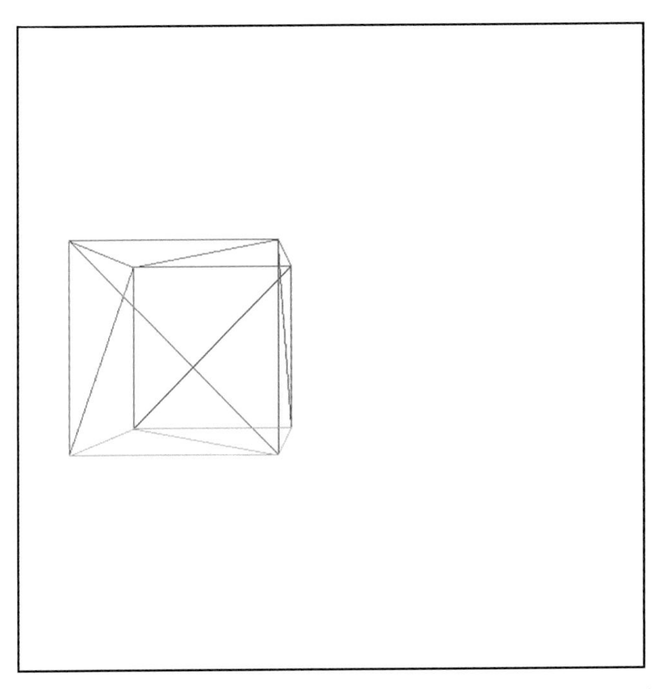

Figure 10-2: Our cube, translated in front of the camera, rendered with wireframe triangles

You can find a live implementation of this algorithm at *https://gabrielgambetta.com/cgfs/scene-demo*.

Models and Instances

What if we want to render two cubes? A naive approach would be to create a new set of vertices and triangles describing a second cube. This would work, but it would waste a lot of memory. What if we wanted to render *one million* cubes?

A better approach is to think in terms of *models* and *instances*. A model is a set of vertices and triangles that describes a certain object in a generic way (think "a cube has eight vertices and six sides"). An instance of a model, on the other hand, describes a concrete occurrence of that model within the scene (think "there's a cube at (0, 0, 5)").

How do we apply this idea in practice? We can have a single description of each unique object in the scene and then place multiple copies of it by specifying their coordinates. Informally, it would be like saying, "This is what a cube looks like, and there's cubes here, here and there."

This is a rough approximation of how we'd describe a scene using this approach:

```
model {
    name = cube
```

```
    vertices {
        ...
    }
    triangles {
        ...
    }
}

instance {
    model = cube
    position = (0, 0, 5)
}

instance {
    model = cube
    position = (1, 2, 3)
}
```

In order to render this, we just go through the list of instances; for each instance, we make a copy of the model's vertices, translate them according to the position of the instance, and then render them as before (Listing 10-2).

```
RenderScene() {
    for I in scene.instances {
        RenderInstance(I);
    }
}

RenderInstance(instance) {
    projected = []
    model = instance.model
    for V in model.vertices {
        V' = V + instance.position
        projected.append(ProjectVertex(V'))
    }
    for T in model.triangles {
        RenderTriangle(T, projected)
    }
}
```

Listing 10-2: An algorithm to render a scene that can contain multiple instances of several objects, each in a different position

If we want this to work as expected, the coordinates of the vertices on the model should be defined in a coordinate system that "makes sense" for the object; we'll call this coordinate system *model space*. For example, we defined our cube such that its center was (0, 0, 0); this means that when we say "a cube located at (1, 2, 3)," we mean "a cube centered around (1, 2, 3)."

After applying the instance translation to the vertices defined in model space, the transformed vertices are now expressed in the coordinate system of the scene; we'll call this coordinate system *world space*.

There are no hard and fast rules to define a model space; it depends on the needs of your application. For example, if you have the model of a person, it might be sensible to place the origin of the coordinate system at their feet.

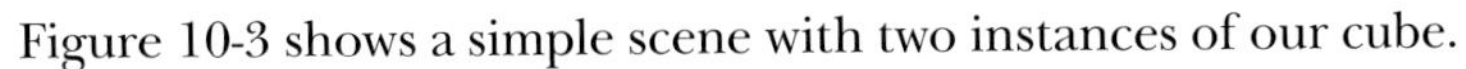

Figure 10-3 shows a simple scene with two instances of our cube.

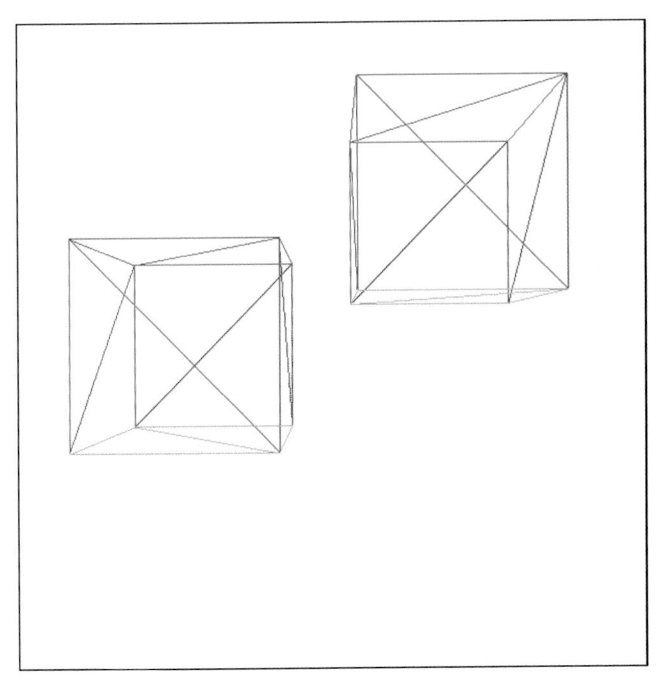

Figure 10-3: A scene with two instances of the same cube model, placed in different positions

You can find a live implementation of this algorithm at *https://gabrielgambetta.com/cgfs/instances-demo*.

Model Transform

The scene definition we described above doesn't give us a lot of flexibility. Since we can only specify the *position* of a cube, we could instantiate as many cubes as we wanted, but they would all be facing the same direction. In general, we want to have more control over the instances: we also want to specify their orientation and possibly their scale.

Conceptually, we can define a *model transform* with these three elements: a scaling factor, a rotation around the origin in model space, and a translation to a specific point in the scene:

```
instance {
    model = cube
    transform {
        scale = 1.5
        rotation = <45 degrees around the Y axis>
        translation = (1, 2, 3)
    }
}
```

We can extend the algorithm in Listing 10-2 to accommodate the new transforms. However, the order in which we *apply* the transforms is important; in particular, the translation must be done last. This is because most of the time we want to rotate and scale the instances around their origin in model space, so we need to do that before they're transformed into world space.

To understand the difference in the results, take a look at Figure 10-4, which shows a 45° rotation around the origin followed by a translation along the Z axis.

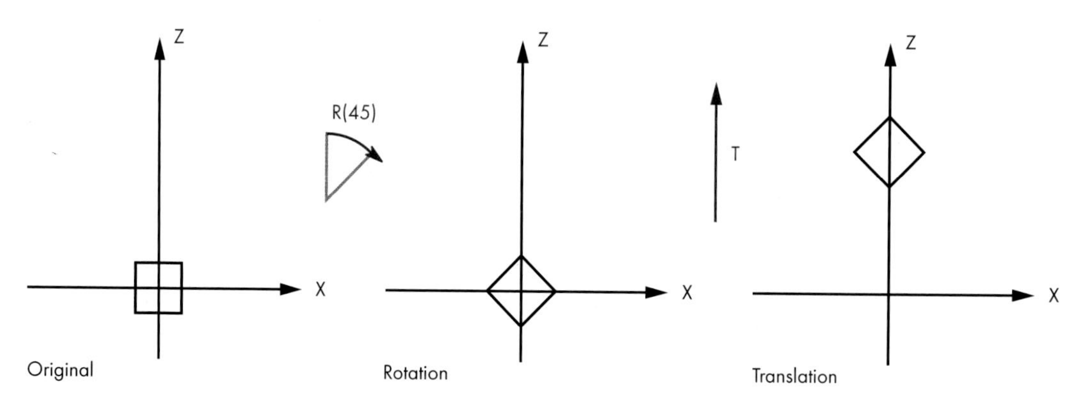

Figure 10-4: Applying rotation and then translation

Figure 10-5 shows the translation applied before the rotation.

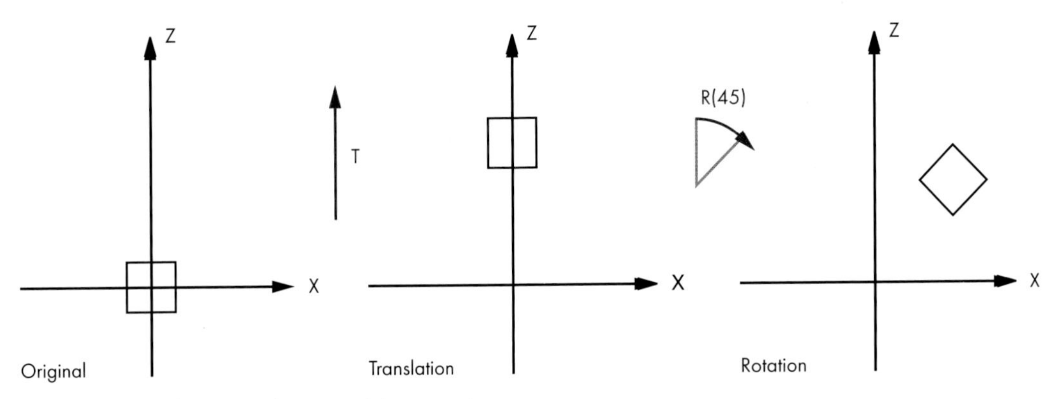

Figure 10-5: Applying translation and then rotation

Strictly speaking, given a rotation followed by a translation, we can find a translation followed by a rotation (perhaps not around the origin) that achieves the same result. However, it's far more natural to express this kind of transform using the first form.

We can write a new version of `RenderInstance` that supports scale, rotation, and position (see Listing 10-3).

```
RenderInstance(instance) {
    projected = []
    model = instance.model
    for V in model.vertices {
        V' = ApplyTransform(V, instance.transform)
        projected.append(ProjectVertex(V'))
    }
    for T in model.triangles {
        RenderTriangle(T, projected)
    }
}
```

Listing 10-3: An algorithm to render a scene that can contain multiple instances of several objects, each with a different transform

The `ApplyTransform` method looks like Listing 10-4.

```
ApplyTransform(vertex, transform) {
    scaled = Scale(vertex, transform.scale)
    rotated = Rotate(scaled, transform.rotation)
    translated = Translate(rotated, transform.translation)
    return translated
}
```

Listing 10-4: A function that applies transforms to a vertex in the correct order

Camera Transform

The previous sections explored how we can position instances of models at different points in the scene. In this section, we'll explore how to move and rotate the camera within the scene.

Imagine you're a camera floating in the middle of a completely empty coordinate system. Suddenly, a red cube appears exactly in front of you (Figure 10-6).

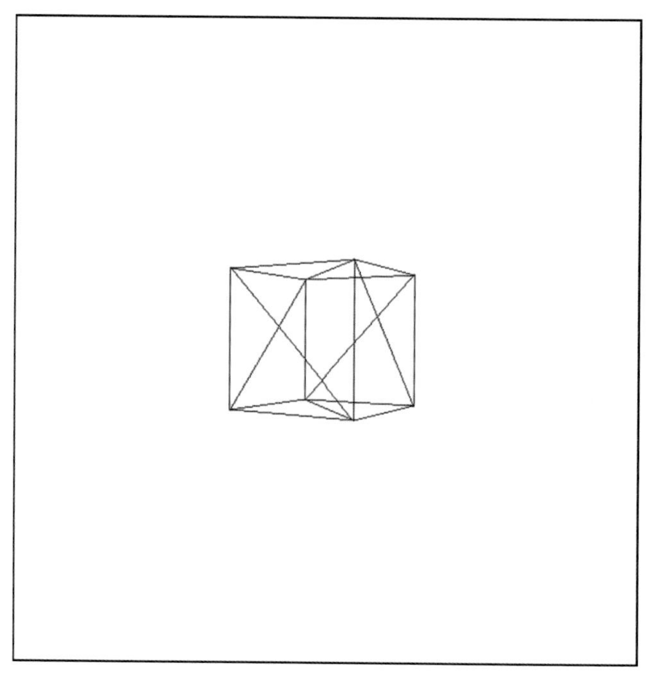

Figure 10-6: A red cube appears in front of the camera.

A second later, the cube moves 1 unit toward you (Figure 10-7).

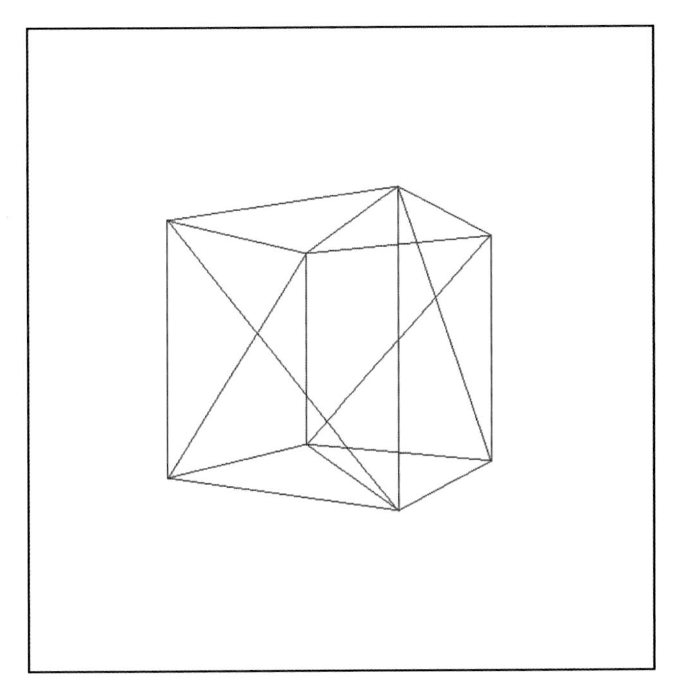

Figure 10-7: The red cube moves toward the camera . . . or does it?

But did the cube really move 1 unit toward you? Or did you move 1 unit toward the cube? Since there are no points of reference at all, and the coordinate system isn't visible, there's no way to tell just by looking at what you see, because the *relative* position of the cube and the camera are identical in both cases (Figure 10-8).

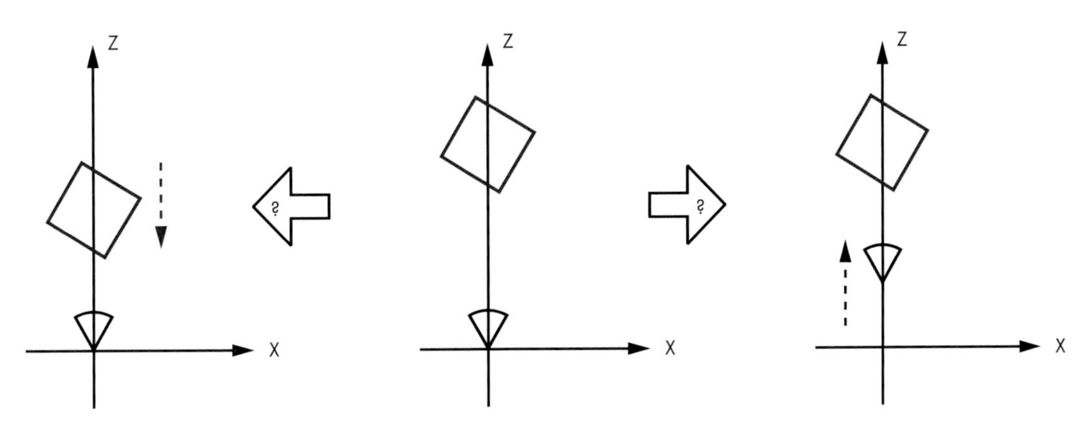

Figure 10-8: Without the coordinate system, we can't tell whether it was the object or the camera that moved.

Now the cube rotates around you 45° clockwise. Or does it? Maybe it was you who rotated 45° counterclockwise? Again, there's no way to tell (Figure 10-9).

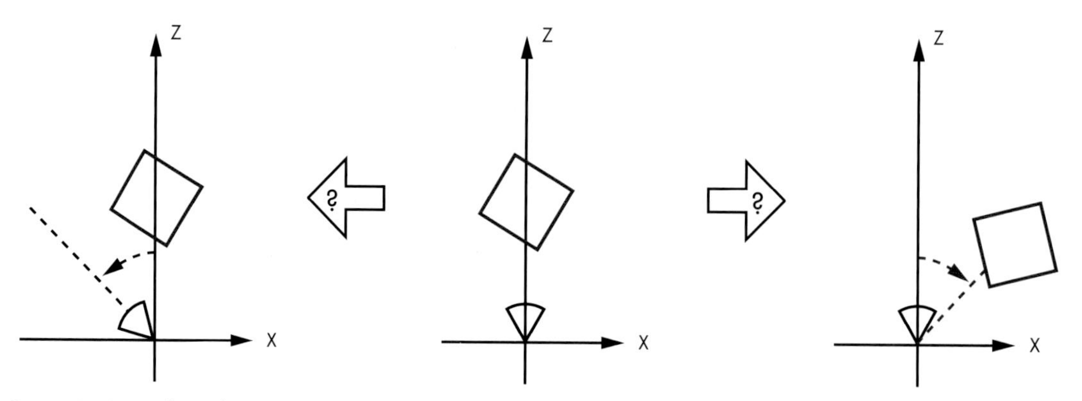

Figure 10-9: Without the coordinate system, we can't tell whether it was the object or the camera that rotated.

What this thought experiment shows is that there's no difference between moving the camera around a fixed scene and keeping the camera fixed while rotating and translating the scene around it!

The advantage of this clearly self-centered vision of the universe is that by keeping the camera fixed at the origin and pointing at $\vec{Z_+}$, we can use the projection equations derived in the previous chapter without any modification. The coordinate system of the camera is called the *camera space*.

Let's assume the camera also has a transform attached to it, consisting of translation and rotation. In order to render the scene from the point of view of the camera, we need to apply the *opposite* transforms to each vertex of the scene:

$$V_{translated} = V_{scene} - camera.translation$$

$$V_{cam\_space} = inverse(camera.rotation) \cdot V_{translated}$$

$$V_{projected} = perspective\_projection(V_{cam\_space})$$

Note that we represent rotations using rotation matrices. Please refer to the Linear Algebra appendix for more details about this.

The Transform Matrix

Now that we can move both the camera and the instances around the scene, let's take a step back and consider everything that happens to a vertex V_{model} in model space until it's projected into the canvas point (cx, cy).

We first apply the model transform to go from model space to world space:

$$V_{model\_scaled} = instance.scale \cdot V_{model}$$

$$V_{model\_rotated} = instance.rotation \cdot V_{model\_scaled}$$

$$V_{world} = V_{model\_rotated} + instance.translation$$

Then we apply the camera transform to go from world space to camera space:

$$V_{translated} = V_{world} - camera.translation$$

$$V_{camera} = inverse(camera.rotation) \cdot V_{translated}$$

Next, we apply the perspective equations to get viewport coordinates:

$$v_x = \frac{V_{camera}x \cdot d}{V_{camera}z}$$

$$v_y = \frac{V_{camera}y \cdot d}{V_{camera}z}$$

And finally we map the viewport coordinates to canvas coordinates:

$$c_x = \frac{v_x \cdot c_w}{v_w}$$

$$c_y = \frac{v_y \cdot c_h}{v_h}$$

As you can see, it's a lot of computation and a lot of intermediate values for each vertex. Wouldn't it be nice if we could reduce all of that to a more compact and efficient form?

Let's express the transforms as functions that take a vertex and return a transformed vertex. Let C_T and C_R be the camera translation and rotation; I_R, I_S, and I_T the instance rotation, scale, and translation; P the perspective projection; and M the viewport-to-canvas mapping. If V is the original vertex and V' is the point on the canvas, we can express all the equations above like this:

$$V' = M(P(C_R^{-1}(C_T^{-1}(I_T(I_R(I_S(V)))))))$$

Ideally, we'd like a single transform F that does whatever the series of original transforms does, but that has a much simpler expression:

$$F = M \cdot P \cdot C_R^{-1} \cdot C_T^{-1} \cdot I_T \cdot I_R \cdot I_S$$

$$V' = F(V)$$

Finding a simple way to represent F isn't trivial. Our main obstacle is that we express each transform in a different way: we express translation as the sum of a point and a vector, rotation as the multiplication of a matrix and a point, scaling as the multiplication of a real number and a point, and perspective projection as real number multiplications and divisions. But if we could express all the transforms in the same way, and if such a way had a mechanism to compose transforms, we'd get the simple transform we want.

Homogeneous Coordinates

Consider the expression $A = (1, 2, 3)$. Does A represent a 3D point or a 3D vector? If we don't know the context in which A is used, there's no way to know.

But let's add a fourth value, called w, to mark A as a point or a vector. If $w = 0$, it's a vector; if $w = 1$, it's a point. So the point A is unambiguously represented as $A = (1, 2, 3, 1)$ and the vector $\vec{A}$ is represented as $(1, 2, 3, 0)$.

Since points and vectors share the same representation, these four-component coordinates are called *homogeneous coordinates*. Homogeneous coordinates have a far deeper and far more involved geometric interpretation, but that's outside the scope of this book; here, we'll just use them as a convenient tool.

Manipulating points and vectors expressed in homogeneous coordinates is compatible with their geometric interpretation. For example, subtracting two points produces a vector:

$$(8, 4, 2, 1) - (3, 2, 1, 1) = (5, 2, 1, 0)$$

Adding two vectors produces another vector:

$$(0, 0, 1, 0) + (1, 0, 0, 0) = (1, 0, 1, 0)$$

In the same way, it's easy to see that adding a point and a vector produces a point, multiplying a vector by a scalar produces a vector, and so on, just as we expect.

So what do coordinates with a w value other than 0 or 1 represent? They also represent points. In fact, any point in 3D has an infinite number of representations in homogeneous coordinates. What matters is the *ratio* between the coordinates and the w value. For example, $(1, 2, 3, 1)$ and $(2, 4, 6, 2)$ represent the same point, as does $(-3, -6, -9, -3)$.

Of all of these representations, we call the one with $w = 1$ the *canonical representation* of the point in homogeneous coordinates; converting any other representation to its canonical representation or to its Cartesian coordinates is trivial:

$$\begin{pmatrix} x \\ y \\ z \\ w \end{pmatrix} = \begin{pmatrix} \frac{x}{w} \\ \frac{y}{w} \\ \frac{z}{w} \\ 1 \end{pmatrix} \rightarrow \begin{pmatrix} \frac{x}{w} \\ \frac{y}{w} \\ \frac{z}{w} \end{pmatrix}$$

So we can convert Cartesian coordinates to homogeneous coordinates, and back to Cartesian coordinates. But how does this help us find a single representation for all the transforms?

Homogeneous Rotation Matrix

Let's begin with a rotation matrix. Converting a 3×3 rotation matrix in Cartesian coordinates to a 4×4 rotation matrix in homogeneous coordinates is trivial; since the w coordinate of the point shouldn't change, we add a column to the right, a row to the bottom, fill them with zeros, and place a 1 in the lower-right element to keep the value of w:

$$\begin{pmatrix} A & B & C \\ D & E & F \\ G & H & I \end{pmatrix} \cdot \begin{pmatrix} x \\ y \\ z \end{pmatrix} = \begin{pmatrix} x' \\ y' \\ z' \end{pmatrix} \rightarrow \begin{pmatrix} A & B & C & 0 \\ D & E & F & 0 \\ G & H & I & 0 \\ 0 & 0 & 0 & 1 \end{pmatrix} \cdot \begin{pmatrix} x \\ y \\ z \\ 1 \end{pmatrix} = \begin{pmatrix} x' \\ y' \\ z' \\ 1 \end{pmatrix}$$

Homogeneous Scale Matrix

A scaling matrix is also trivial in homogeneous coordinates, and it's constructed in the same way as the rotation matrix:

$$\begin{pmatrix} S_x & 0 & 0 \\ 0 & S_y & 0 \\ 0 & 0 & S_z \end{pmatrix} \cdot \begin{pmatrix} x \\ y \\ z \end{pmatrix} = \begin{pmatrix} x \cdot S_x \\ y \cdot S_y \\ z \cdot S_z \end{pmatrix} \rightarrow \begin{pmatrix} S_x & 0 & 0 & 0 \\ 0 & S_y & 0 & 0 \\ 0 & 0 & S_z & 0 \\ 0 & 0 & 0 & 1 \end{pmatrix} \cdot \begin{pmatrix} x \\ y \\ z \\ 1 \end{pmatrix} = \begin{pmatrix} x \cdot S_x \\ y \cdot S_y \\ z \cdot S_z \\ 1 \end{pmatrix}$$

Homogeneous Translation Matrix

The rotation and scale matrices were easy; they were already expressed as matrix multiplications in Cartesian coordinates, we just had to add a 1 to preserve the w coordinate. But what can we do with a translation, which we had expressed as an addition in Cartesian coordinates?

We're looking for a 4×4 matrix such that

$$\begin{pmatrix} T_x \\ T_y \\ T_z \\ 0 \end{pmatrix} + \begin{pmatrix} x \\ y \\ z \\ 1 \end{pmatrix} = \begin{pmatrix} A & B & C & D \\ E & F & G & H \\ I & J & K & L \\ M & N & O & P \end{pmatrix} \cdot \begin{pmatrix} x \\ y \\ z \\ 1 \end{pmatrix} = \begin{pmatrix} x + T_x \\ y + T_y \\ z + T_z \\ 1 \end{pmatrix}$$

Let's focus on getting $x + T_x$ first. This value is the result of multiplying the first row of the matrix and the point—that is,

$$\begin{pmatrix} A & B & C & D \end{pmatrix} \cdot \begin{pmatrix} x \\ y \\ z \\ 1 \end{pmatrix} = x + T_x$$

If we expand the vector multiplication, we get

$$Ax + By + Cz + D = x + T_x$$

From here we can deduce that $A = 1$, $B = C = 0$, and $D = T_x$.

Following a similar reasoning for the rest of the coordinates, we arrive at the following matrix expression for the translation:

$$\begin{pmatrix} T_x \\ T_y \\ T_z \\ 0 \end{pmatrix} + \begin{pmatrix} x \\ y \\ z \\ 1 \end{pmatrix} = \begin{pmatrix} 1 & 0 & 0 & T_x \\ 0 & 1 & 0 & T_y \\ 0 & 0 & 1 & T_z \\ 0 & 0 & 0 & 1 \end{pmatrix} \cdot \begin{pmatrix} x \\ y \\ z \\ 1 \end{pmatrix} = \begin{pmatrix} x + T_x \\ y + T_y \\ z + T_z \\ 1 \end{pmatrix}$$

Homogeneous Projection Matrix

Sums and multiplications are easy to express as multiplications of matrices and vectors because they involve, after all, sums and multiplications. But the perspective projection equations have a division by z. How can we express that?

You may be tempted to think that dividing by z is the same as multiplying by $1/z$, and you may want to solve this problem by putting $1/z$ in the matrix. However, *which* z coordinate would we put there? We want this projection matrix to work for *every* input point, so hardcoding the z coordinate of *any* point would not give us what we want.

Fortunately, homogeneous coordinates do have one instance of a division: the division by the w coordinate when converting back to Cartesian coordinates. If we can manage to make the z coordinate of the original point appear as the w coordinate of the "projected" point, we'll get the projected x and y once we convert the point back to Cartesian coordinates:

$$\begin{pmatrix} A & B & C & D \\ E & F & G & H \\ I & J & K & L \end{pmatrix} \cdot \begin{pmatrix} x \\ y \\ z \\ 1 \end{pmatrix} = \begin{pmatrix} x \cdot d \\ y \cdot d \\ z \end{pmatrix} \rightarrow \begin{pmatrix} \frac{x \cdot d}{z} \\ \frac{y \cdot d}{z} \end{pmatrix}$$

Note that this matrix is 3×4; it can be multiplied by a four-element vector (the transformed 3D point in homogeneous coordinates) and it will yield a three-element vector (the projected 2D point in homogeneous coordinates), which is then converted to 2D Cartesian coordinates by dividing by w. This gives us exactly the values of x' and y' we were looking for. The missing element here is z', which we know is equal to d by definition.

Applying the same reasoning we used to deduce the translation matrix, we can express the perspective projection as follows:

$$\begin{pmatrix} d & 0 & 0 & 0 \\ 0 & d & 0 & 0 \\ 0 & 0 & 1 & 0 \end{pmatrix} \cdot \begin{pmatrix} x \\ y \\ z \\ 1 \end{pmatrix} = \begin{pmatrix} x \cdot d \\ y \cdot d \\ z \end{pmatrix} \rightarrow \begin{pmatrix} \frac{x \cdot d}{z} \\ \frac{y \cdot d}{z} \end{pmatrix}$$

Homogeneous Viewport-to-Canvas Matrix

The last step is mapping the projected point on the viewport to the canvas. This is just a 2D scaling transform with $S_x = \frac{c_w}{v_w}$ and $S_y = \frac{c_h}{v_h}$. This matrix is thus

$$\begin{pmatrix} \frac{c_w}{v_w} & 0 & 0 \\ 0 & \frac{c_w}{v_w} & 0 \\ 0 & 0 & 1 \end{pmatrix} \cdot \begin{pmatrix} x \\ y \\ z \end{pmatrix} = \begin{pmatrix} \frac{x \cdot c_w}{v_w} \\ \frac{y \cdot c_h}{v_h} \\ z \end{pmatrix}$$

In fact, it's easy to combine this with the projection matrix to get a simple 3D-to-canvas matrix:

$$\begin{pmatrix} \frac{d \cdot cw}{vw} & 0 & 0 & 0 \\ 0 & \frac{d \cdot ch}{vh} & 0 & 0 \\ 0 & 0 & 1 & 0 \end{pmatrix} \cdot \begin{pmatrix} x \\ y \\ z \\ 1 \end{pmatrix} = \begin{pmatrix} \frac{x \cdot d \cdot cw}{vw} \\ \frac{y \cdot d \cdot cw}{vh} \\ z \end{pmatrix} \rightarrow \begin{pmatrix} (\frac{x \cdot d}{z})(\frac{cw}{vw}) \\ (\frac{y \cdot d}{z})(\frac{ch}{vh}) \end{pmatrix}$$

The Transform Matrix Revisited

After all this work, we can express every transform we need to convert a model vertex V into a canvas pixel V' as a matrix. Moreover, we can compose these transforms by multiplying their corresponding matrices. So we can express the whole sequence of transforms as a single matrix:

$$F = M \cdot P \cdot C_R^{-1} \cdot C_T^{-1} \cdot I_T \cdot I_R \cdot I_S$$

Now transforming a vertex is just a matter of computing the following matrix-by-point multiplication:

$$V' = F \cdot V$$

Furthermore, we can decompose the transform into three parts:

$$M_{Projection} = M \cdot P$$

$$M_{Camera} = C_R^{-1} \cdot C_T^{-1}$$

$$M_{Model} = I_T \cdot I_R \cdot I_S$$

$$M = M_{Projection} \cdot M_{Camera} \cdot M_{Model}$$

These matrices don't need to be computed from scratch for every vertex (that's the point of using a matrix after all). Because matrix multiplication is associative, we can reuse the parts of the expression that don't change.

$M_{Projection}$ should rarely change; it only depends on the size of the viewport and the size of the canvas. The size of the canvas changes when, for example, the application goes from windowed to fullscreen. The size of the viewport would only change if the field of view of the camera changes; this doesn't happen very often.

M_{Camera} may change every frame; it depends on the camera position and orientation, so if the camera is moving or turning, it needs to be recomputed. Once computed, though, it remains constant for every object drawn in the frame, so it would be computed at most once per frame.

M_{Model} will be different for each instance in the scene; however, it will remain constant over time for instances that don't move (for example, trees and buildings), so it can be computed once and stored in the scene itself. For objects that do move (for example, cars in a racing game) it needs to be computed every time they move (which is likely to be every frame).

A very high level of the scene rendering pseudocode would look like Listing 10-5.

```
RenderModel(model, transform) {
    projected = []
    for V in model.vertices {
        projected.append(ProjectVertex(transform * V))
    }
    for T in model.triangles {
        RenderTriangle(T, projected)
    }
}


RenderScene() {
    M_camera = MakeCameraMatrix(camera.position, camera.orientation)

    for I in scene.instances {
        M = M_camera * I.transform
```

```
        RenderModel(I.model, M)
    }
}
```

Listing 10-5: An algorithm to render a scene using transform matrices

We can now draw a scene containing several instances of different models, possibly moving around and rotating, and we can move the camera throughout the scene. Figure 10-10 shows two instances of our cube model, each with a different transform (including translation and rotation), rendered from a translated and rotated camera.

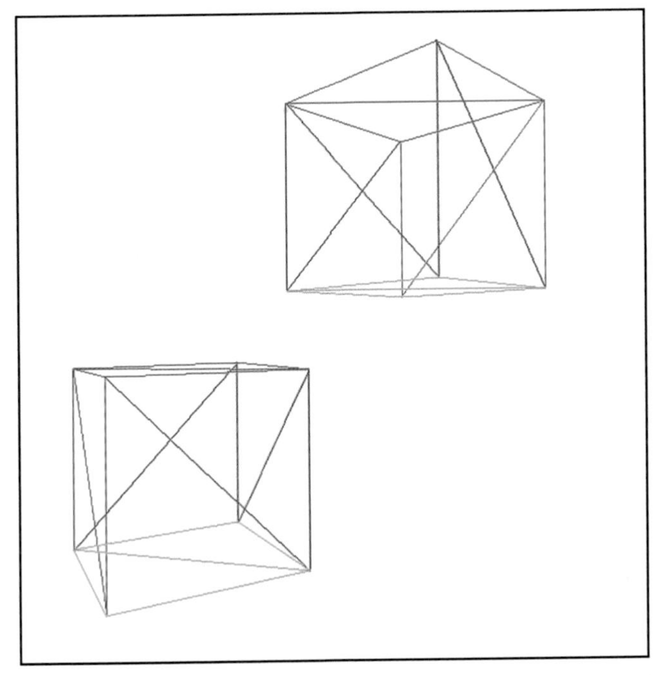

Figure 10-10: A scene with two instances of the same cube model, having different instance transforms, and a transformed camera

You can find a live implementation of this algorithm at *https://gabrielgambetta.com/cgfs/transforms-demo*.

Summary

We covered a lot of ground in this chapter. We first explored how to represent models made out of triangles. Then we figured out how to apply the perspective projection equation we derived in the previous chapter to entire models, so we can go from an abstract 3D model to its representation on the screen.

Next we developed a way to have multiple instances of the same model in the scene without having multiple copies of the model itself. Then we found out how to lift one of the limitations we had been working with so far: our camera no longer needs to be fixed at the origin of the coordinate system or pointing toward $\vec{Z}+$.

Finally, we explored how to represent all the transforms we need to apply to a vertex as matrix multiplications in homogeneous coordinates, and this allowed us to reduce the computations required to render a scene by condensing many of the consecutive transforms into just three matrices: one for the perspective projection and viewport-to-canvas mapping, one for the instance transform, and one for the camera transform.

This has given us a lot of flexibility in terms of what we can represent in a scene, and it also allows us to move the camera around the scene. But we still have two important limitations. First, moving the camera means we can end up with objects behind it, which causes all sorts of problems. Second, the rendering doesn't look so great: it's still a wireframe image.

Note that for practical reasons we won't be using the full projection matrix in the rest of this book. Instead, we'll use the model and camera transforms separately and then convert their results back to Cartesian coordinates as follows:

$$x' = \frac{x \cdot d \cdot cw}{z \cdot vw}$$

$$y' = \frac{y \cdot d \cdot ch}{z \cdot vh}$$

This lets us do some more operations in 3D that can't be expressed as matrix transforms before we project the points.

In the next chapter, we'll deal with objects that shouldn't be visible, and then we'll spend the rest of this book making the rendered objects look better.

11

CLIPPING

In the last few chapters, we developed equations and algorithms to transform a 3D definition of a scene into 2D shapes we can draw on the canvas; we developed a scene structure that lets us define 3D models and place instances of those models in the scene; and we developed an algorithm that lets us render the scene from any point of view.

However, doing this exposes one of the limitations we've been working with: the perspective projection equations only work as expected for points that are in front of the camera. Since we can now move and rotate the camera around the scene, this poses a problem.

In this chapter, we'll develop the techniques necessary to lift this limitation: we'll explore how to identify points, triangles, and entire objects that are behind the camera and develop techniques to deal with them.

An Overview of the Clipping Process

Back in Chapter 9, we arrived at the following equations:

$$P'_x = \frac{P_x \cdot d}{P_z}$$

$$P'_y = \frac{P_y \cdot d}{P_z}$$

The division by P_z is problematic; it can cause a division by zero. Moreover, points behind the camera have negative Z values, which we currently can't handle properly. Even points in front of the camera but very close to it will cause trouble in the form of severely distorted objects.

To avoid these problematic cases, we'll choose not to render anything behind the projection plane $Z = d$. This *clipping plane* lets us classify any point as being *inside* or *outside* of the *clipping volume*—that is, the subset of space that is actually visible from the camera. In this case, the clipping volume is "whatever is in front of $Z = d$." We'll only render the parts of the scene that are inside the clipping volume.

The Clipping Volume

Using a single clipping plane to make sure no objects behind the camera are rendered will produce correct results, but it's not entirely efficient. Some objects may be in front of the camera but still not visible; for example, the projection of an object near the projection plane but far, far to the right will be projected outside of the viewport and therefore won't be visible, as shown in Figure 11-1.

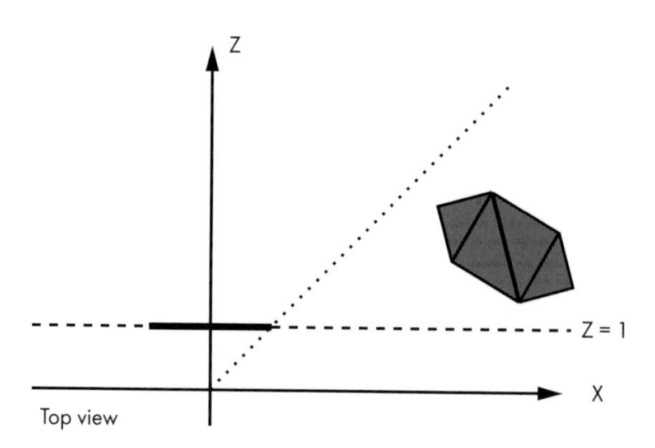

Figure 11-1: An object that is in front of the projection plane, but will be projected outside of the viewport

Any computational resources we use to project such an object, plus all the per-triangle and per-vertex computations done to render it, would be wasted. It would be more efficient to ignore these objects altogether.

To do this, we can define additional planes to clip the scene to *exactly* what should be visible on the viewport; these planes are defined by the camera and each of the four sides of the viewport (Figure 11-2).

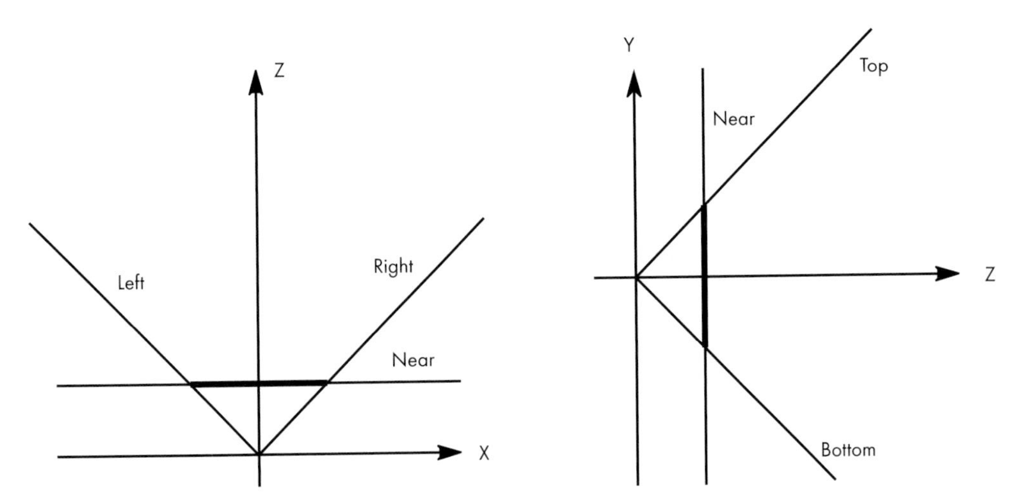

Figure 11-2: The five planes that define our clipping volume

Each of the clipping planes splits space in two parts we call *half-spaces*. The "inside" half-space is everything that's in front of the plane; the "outside" half-space is everything that's behind it. The "inside" of the clipping volume we're defining is the *intersection* of the "inside" half-spaces defined by each clipping plane. In this case, the clipping volume looks like an infinitely tall pyramid with the top chopped off.

This means that to clip the scene against a clipping volume, we just need to clip it in succession against each of the planes that define the clipping volume. Whatever geometry remains inside after clipping against one plane is then clipped against the remaining planes. After the scene has been clipped against all the planes, the geometry that remains is the result of clipping the scene against the clipping volume.

Next we'll take a look at how to clip the scene against each clipping plane.

Clipping the Scene Against a Plane

Consider a scene with multiple objects, each made of four triangles (Figure 11-3).

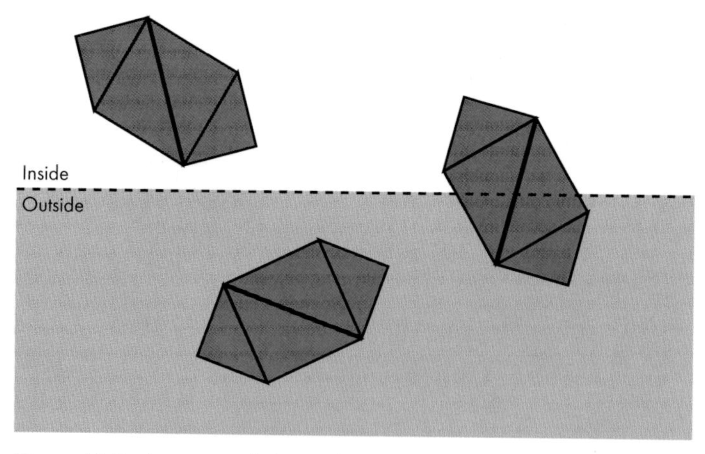

Figure 11-3: A scene with three objects

The fewer operations we execute, the faster our renderer will be. We will clip the scene against a clipping plane as a sequence of stages. Each stage will attempt to classify as much geometry as possible as either *accepted* or *discarded*, depending on whether it's inside or outside the half-space defined by the clipping plane (that is, the clipping volume of this plane). Whatever geometry can't be classified moves on to the next stage, which will take a more detailed look at it.

The first stage attempts to classify entire objects at once. If an object is completely inside the clipping volume, it's accepted (green in Figure 11-4); if it's completely outside, it's discarded (red in Figure 11-4).

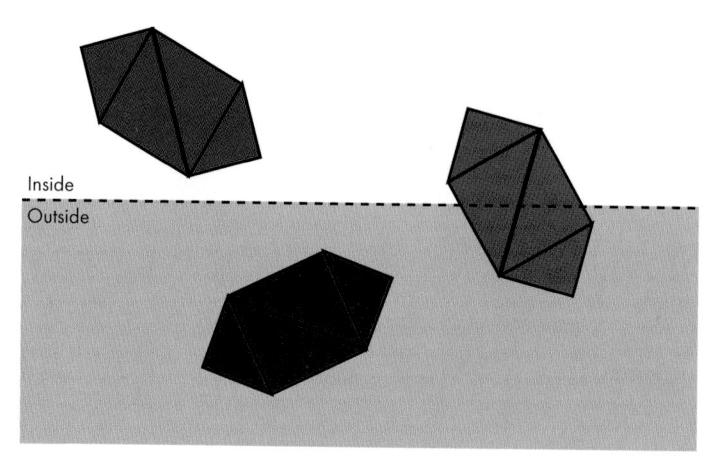

Figure 11-4: Clipping at the object level. Green is accepted, red is discarded, and gray requires further processing.

If an object can't be fully accepted or discarded, we move on to the next stage and classify each of its triangles independently. If a triangle is completely inside the clipping volume, it's accepted; if it's completely outside, it's discarded (see Figure 11-5).

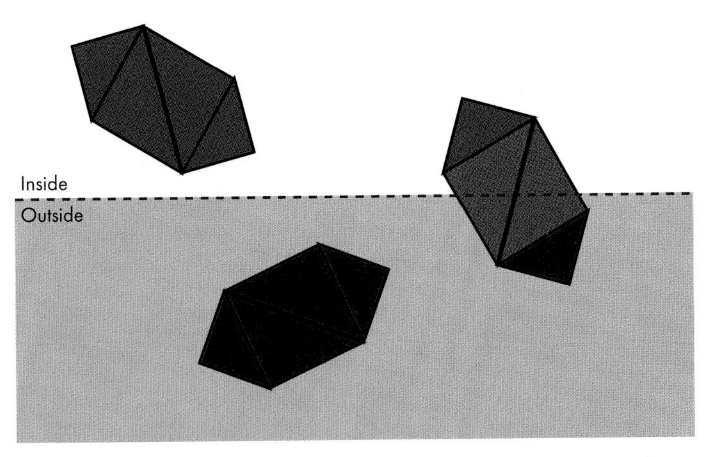

Figure 11-5: Clipping at the triangle level. Each triangle of the rightmost object is either accepted, discarded, or requires further processing.

Finally, for each triangle that isn't either accepted or discarded, we need to clip the triangle itself. The original triangle is removed, and either one or two new triangles are added to cover the part of the triangle that is inside the clipping volume (see Figure 11-6).

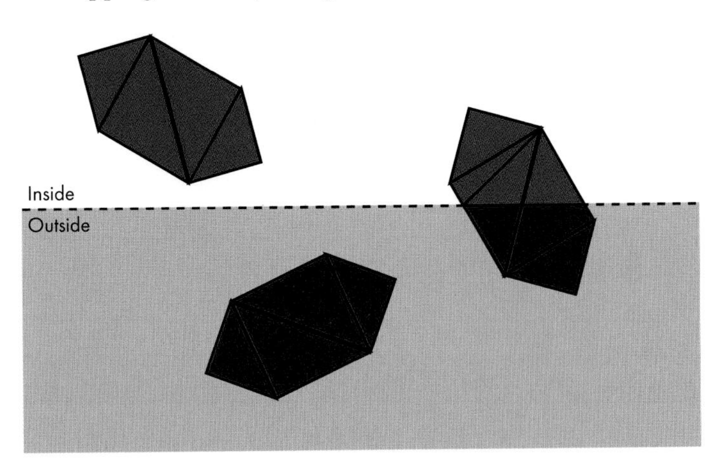

Figure 11-6: Clipping at the vertex level. Each triangle that is partially inside the clipping volume is split into one or two triangles that are fully inside the clipping volume.

Now that we have a clear conceptual understanding of how clipping works, we'll develop the math and algorithms to create a working implementation.

Defining the Clipping Planes

Let's start with the equation of the projection plane $Z = d$, which we'll use as a clipping plane. This equation is simple to visualize, but it's not in the most convenient or general form for our purposes.

The general equation for a 3D plane is $Ax + By + Cz + D = 0$, meaning a point $P = (x, y, z)$ will satisfy that equation if and only if P is on the plane. If we group the coefficients (A, B, C) in a vector $\vec{N}$, we can rewrite the equation as $\langle \vec{N}, P \rangle + D = 0$.

Note that if $\langle \vec{N}, P \rangle + D = 0$, then $k\langle \vec{N}, P \rangle + kD = 0$ for any value of k. In particular, we can choose $k = 1/|\vec{N}|$, multiply the original equation, and get a new equation $\langle \vec{N}', P \rangle + D' = 0$ where $\vec{N}'$ is a unit vector. So any given plane can be represented by an equation $\langle \vec{N}, P \rangle + D = 0$, where $\vec{N}$ is a unit vector and D is a real number.

This is a very convenient formulation: $\vec{N}$ happens to be the normal of the plane and $-D$ is the *signed distance* from the origin to the plane. In fact, for any point P, $\langle \vec{N}, P \rangle + D$ is the signed distance from the plane to P; *distance* = 0 is just the special case where P is contained in the plane.

If $\vec{N}$ is the normal of a plane, so is $-\vec{N}$, so we choose $\vec{N}$ such that it points to "inside" the clipping volume. For the plane $Z = d$, we choose the normal $(0, 0, 1)$, which points "forward" with respect to the camera. Since the point $(0, 0, d)$ is contained in the plane, it must satisfy the plane equation, and we can solve for D:

$$\langle \vec{N}, P \rangle + D = \langle (0, 0, 1), (0, 0, d) \rangle + D = d + D = 0$$

and from this we immediately get $D = -d$.

We could have gotten $D = -d$ directly from the original plane equation $Z = d$ by rewriting it as $Z - d = 0$. However, we can apply this general method to derive the equations of the rest of the clipping planes.

We know all these additional planes have $D = 0$ (because they all go through the origin), so all we need to do is determine their normals. To make the math simple, we'll choose a 90° field of view (FOV), meaning the planes are at 45°.

Consider the left clipping plane. The direction of its normal is $(1, 0, 1)$ (that is, 45° right and forward). The length of that vector is $\sqrt{2}$, so if we normalize it we get $(\frac{1}{\sqrt{2}}, 0, \frac{1}{\sqrt{2}})$. Therefore the equation of the left clipping plane is

$$\langle N, P \rangle + D = \langle (\frac{1}{\sqrt{2}}, 0, \frac{1}{\sqrt{2}}), P \rangle = 0$$

Similarly, the normals for the right, bottom, and top clipping planes are $(\frac{-1}{\sqrt{2}}, 0, \frac{1}{\sqrt{2}})$, $(0, \frac{1}{\sqrt{2}}, \frac{1}{\sqrt{2}})$, and $(0, \frac{-1}{\sqrt{2}}, \frac{1}{\sqrt{2}})$ respectively. Computing the clipping planes for any arbitrary FOV would involve a little bit of trigonometry.

In summary, our clipping volume is defined by the following five planes:

$$(near) \quad \langle (0, 0, 1), P \rangle - d = 0$$

$$(left) \quad \langle (\frac{1}{\sqrt{2}}, 0, \frac{1}{\sqrt{2}}), P \rangle = 0$$

$$(right) \quad \langle (\frac{-1}{\sqrt{2}}, 0, \frac{1}{\sqrt{2}}), P \rangle = 0$$

$$(bottom) \quad \langle (0, \frac{1}{\sqrt{2}}, \frac{1}{\sqrt{2}}), P \rangle = 0$$

$$(top) \quad \langle (0, \frac{-1}{\sqrt{2}}, \frac{1}{\sqrt{2}}), P \rangle = 0$$

Let's now take a detailed look at how to clip geometry against a plane.

Clipping Whole Objects

Suppose we put each model inside the smallest sphere that can contain it; we call that sphere the *bounding sphere* of the object. Computing this sphere is surprisingly more difficult than it seems, and it falls outside the scope of this book. But a rough approximation can be obtained by first computing the center of the sphere by averaging the coordinates of all the vertices in the model, and then defining the radius to be the distance from the center to the vertex that it's farthest away from.

In any case, let's assume we know the center C and the radius r of a sphere that completely contains each model. Figure 11-7 shows a scene with a few objects and their bounding spheres.

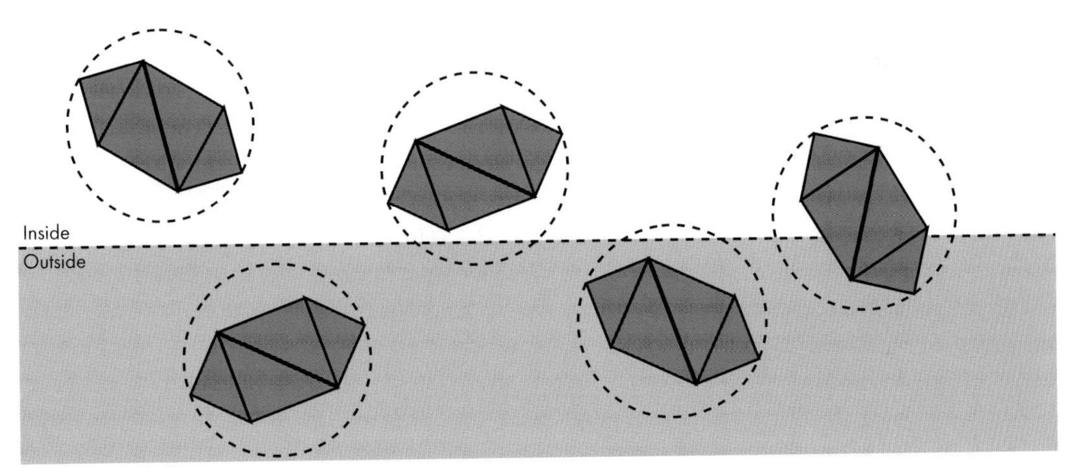

Figure 11-7: A scene with a few objects and their bounding spheres

We can categorize the spatial relationship between this sphere and a plane as follows:

The sphere is completely in front of the plane. In this case, the entire object is accepted; no further clipping is necessary against this plane (but it may still be clipped by a different plane). See Figure 11-8 for an example.

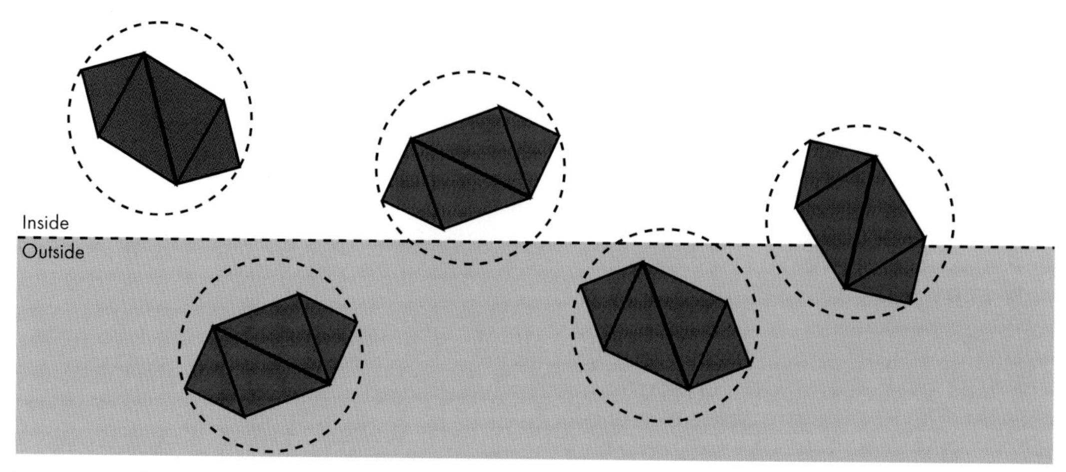

Figure 11-8: The green object is accepted.

The sphere is completely behind the plane. In this case, the entire object is discarded; no further clipping is necessary (no matter what the other planes are, no part of the object will ever be inside the clipping volume). See Figure 11-9 for an example.

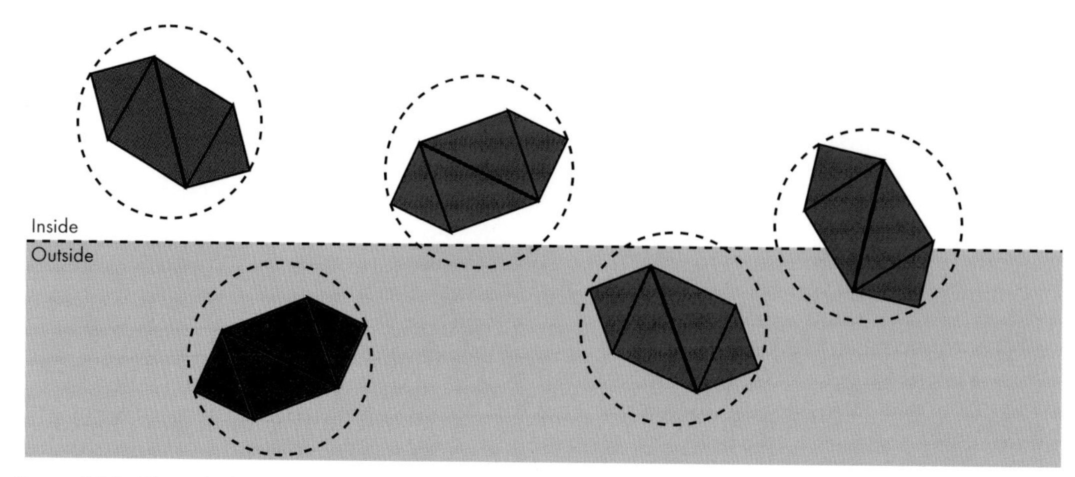

Figure 11-9: The red object is discarded.

The plane intersects the sphere. This doesn't give us enough information to know whether any part of the object is inside the clipping volume; it may be completely inside, completely outside, or partially inside. It is necessary to proceed to the next step and clip the model triangle by triangle. See Figure 11-10 for an example.

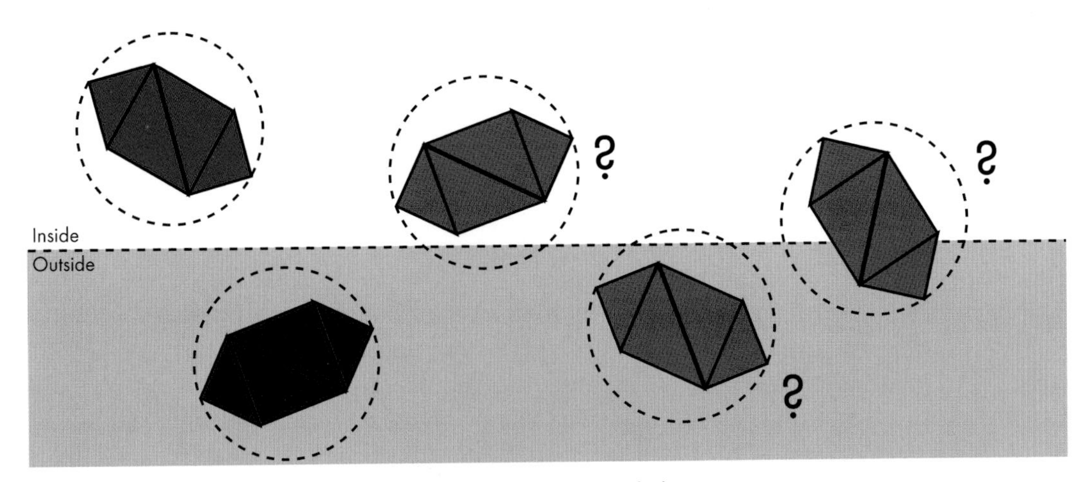

Figure 11-10: The gray objects can't be fully accepted or discarded.

How does this categorization actually work? The way we've chosen to express the clipping planes is such that plugging any point into the plane equation gives us the signed distance from the point to the plane; in particular, we can compute the signed distance d from the center of the bounding sphere to the plane. So if $d > r$, the sphere is in front of the plane; if $d < -r$, the sphere is behind the plane; otherwise $|d| < r$, which means the plane intersects the sphere. Figure 11-11 illustrates all three cases.

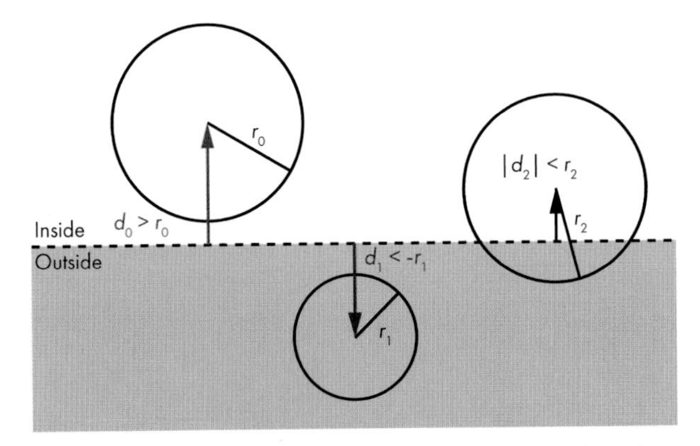

Figure 11-11: The signed distance from the center of a sphere to a clipping plane tells us whether the sphere is in front of the plane, behind the plane, or intersects the plane.

Clipping Triangles

If the sphere–plane test isn't enough to determine whether an object is fully in front or fully behind the clipping plane, we have to clip each triangle against it.

We can classify each vertex of the triangle against the clipping plane by looking at its signed distance to the plane. If the distance is zero or positive, the vertex is in front of the clipping plane; otherwise, it's behind. Figure 11-12 illustrates this idea.

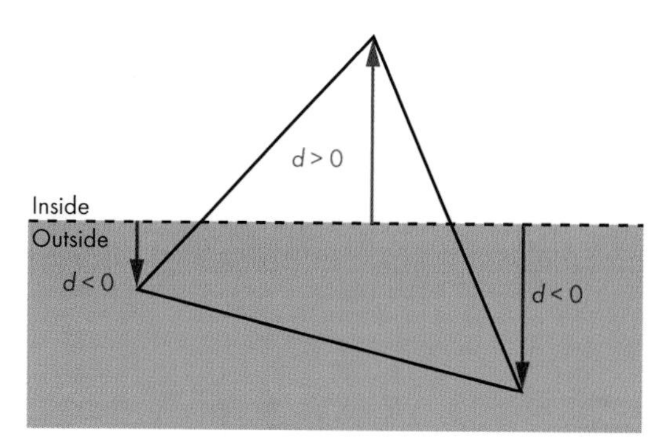

Figure 11-12: The signed distance from a vertex to a clipping plane tells us whether the vertex is in front of or behind the plane.

For each triangle, there are four possible classifications:

Three vertices in front. In this case, the whole triangle is in front of the clipping plane, so we accept it and no further clipping against this plane is needed.

Three vertices behind. In this case, the whole triangle is behind the clipping plane, so we discard it and no further clipping is necessary at all.

One vertex in front. Let A be the vertex of the triangle ABC that is in front of the plane. In this case, we discard ABC, and add a new triangle $AB'C'$, where B' and C' are the intersections of AB and AC with the clipping plane (Figure 11-13).

Two vertices in front. Let A and B be the vertices of the triangle ABC that are in front of the plane. In this case, we discard ABC and add two new triangles: ABA' and $A'BB'$, where A' and B' are the intersections of AC and BC with the clipping plane (Figure 11-14).

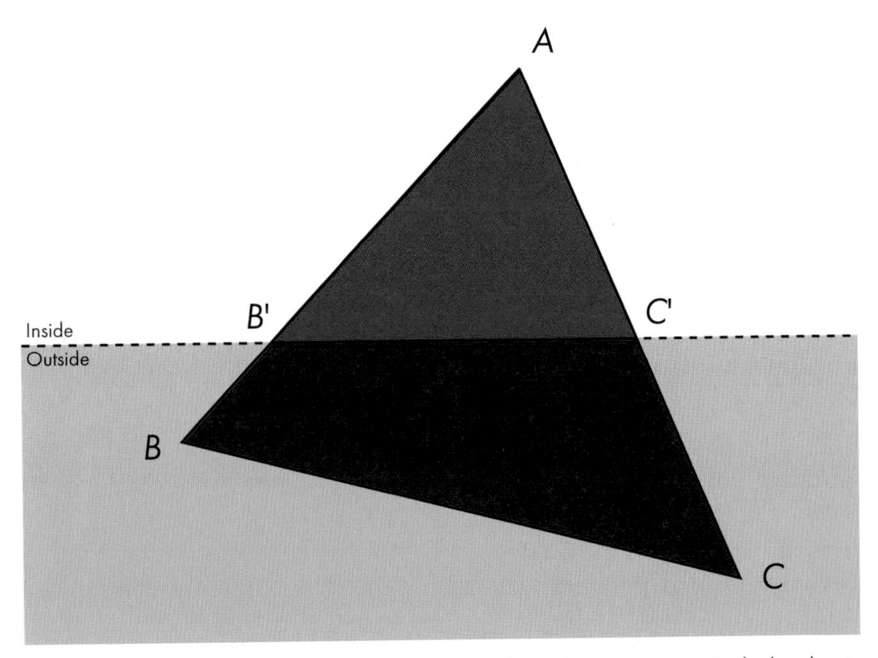

Figure 11-13: A triangle ABC *with one vertex inside and two vertices outside the clipping volume is replaced by a single triangle* AB′C′*.*

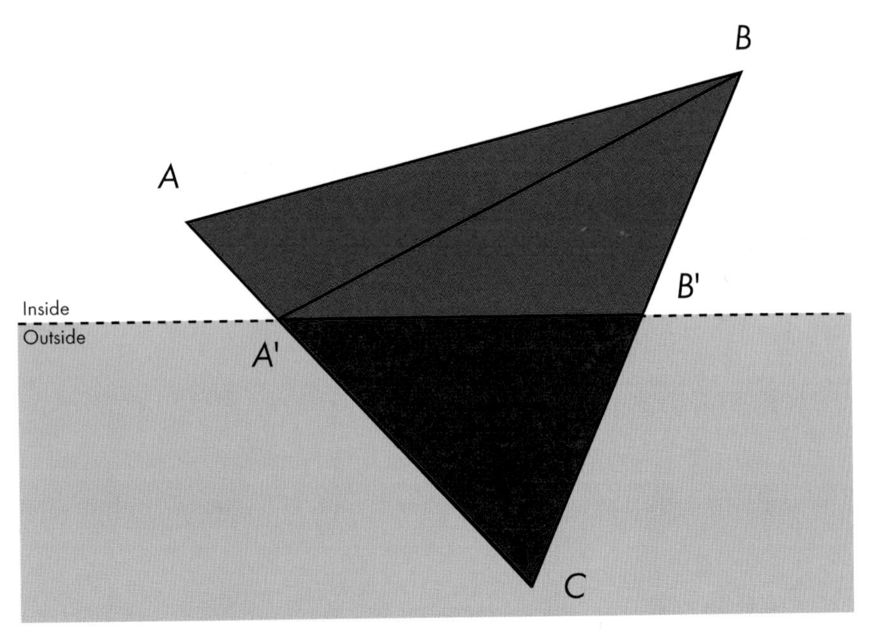

Figure 11-14: A triangle ABC *with one vertex outside and two vertices inside the clipping volume is replaced by two triangles* ABA′ *and* A′BB′*.*

Segment-Plane Intersection

To clip triangles as discussed above, we need to compute the intersection of the sides of the triangle with the clipping plane.

We have a clipping plane given by the equation $\langle N, P \rangle + D = 0$. The triangle side AB can be expressed with a parametric equation as $P = A + t(B - A)$ for $0 \leq t \leq 1$. To compute the value of the parameter t where the intersection occurs, we replace P in the plane equation with the parametric equation of the segment:

$$\langle N, P \rangle + D = 0$$

$$P = A + t(B - A)$$

$$\Longrightarrow \langle N, A + t(B - A) \rangle + D = 0$$

Using the linear properties of the dot product:

$$\langle N, A \rangle + t\langle N, B - A \rangle + D = 0$$

Solving for t:

$$t = \frac{-D - \langle N, A \rangle}{\langle N, B - A \rangle}$$

We know a solution always exists because we know AB intersects the plane; mathematically, $\langle N, B - A \rangle$ can't be zero because that would imply that the segment and the normal are perpendicular, which in turn would imply that the segment and the plane don't intersect.

Having computed t, the intersection Q is simply

$$Q = A + t(B - A)$$

Note that if the original vertices carry additional attributes (for example, the h intensity value we were using in Chapter 7), we need to compute the values of these attributes for the new vertices.

In the equation above, t is the fraction of the segment AB where the intersection occurs. Let α_A and α_B be the values of some attribute α at the points A and B; if we assume the attribute varies linearly across AB, then α_Q can be computed as

$$\alpha_Q = \alpha_A + t(\alpha_B - \alpha_A)$$

We now have all the algorithms and equations to implement our clipping pipeline.

Clipping Pseudocode

Let's write some high-level pseudocode for the clipping pipeline. We'll follow the top-down approach we developed before.

To clip a scene, we clip each of its instances (Listing 11-1).

```
ClipScene(scene, planes) {
    clipped_instances = []
    for I in scene.instances {
        clipped_instance = ClipInstance(I, planes)
        if clipped_instance != NULL {
            clipped_instances.append(clipped_instance)
        }
    }
    clipped_scene = Copy(scene)
    clipped_scene.instances = clipped_instances
    return clipped_scene
}
```

Listing 11-1: An algorithm to clip a scene against a set of clipping planes

To clip an instance, we either accept it, reject it, or clip each of its triangles, depending on its bounding sphere (Listing 11-2).

```
ClipInstance(instance, planes) {
    for P in planes {
        instance = ClipInstanceAgainstPlane(instance, plane)
        if instance == NULL {
            return NULL
        }
    }
    return instance
}

ClipInstanceAgainstPlane(instance, plane) {
    d = SignedDistance(plane, instance.bounding_sphere.center)
    if d > r {
        return instance
    } else if d < -r {
        return NULL
    } else {
        clipped_instance = Copy(instance)
        clipped_instance.triangles =
            ClipTrianglesAgainstPlane(instance.triangles, plane)
```

```
        return clipped_instance
    }
}
```

Listing 11-2: An algorithm to clip an instance against a set of clipping planes

Finally, to clip a triangle, we either accept it, reject it, or decompose it into up to two triangles, depending on how many of its vertices are in front of the clipping plane (Listing 11-3).

```
ClipTrianglesAgainstPlane(triangles, plane) {
    clipped_triangles = []
    for T in triangles {
        clipped_triangles.append(ClipTriangle(T, plane))
    }
    return clipped_triangles
}


ClipTriangle(triangle, plane) {
    d0 = SignedDistance(plane, triangle.v0)
    d1 = SignedDistance(plane, triangle.v1)
    d2 = SignedDistance(plane, triangle.v2)

    if {d0, d1, d2} are all positive {
        return [triangle]
    } else if {d0, d1, d2} are all negative {
        return []
    } else if only one of {d0, d1, d2} is positive {
        let A be the vertex with a positive distance
        compute B' = Intersection(AB, plane)
        compute C' = Intersection(AC, plane)
        return [Triangle(A, B', C')]
    } else /* only one of {d0, d1, d2} is negative */ {
        let C be the vertex with a negative distance
        compute A' = Intersection(AC, plane)
        compute B' = Intersection(BC, plane)
        return [Triangle(A, B, A'), Triangle(A', B, B')]
    }
}
```

Listing 11-3: An algorithm to clip a set of triangles against a clipping plane

The helper function `SignedDistance` just plugs the coordinates of a point into the equation of a plane (Listing 11-4).

```
SignedDistance(plane, vertex) {
    normal = plane.normal
    return (vertex.x * normal.x)
         + (vertex.y * normal.y)
         + (vertex.z * normal.z)
         + plane.D
}
```

Listing 11-4: A function to compute the signed distance from a plane to a point

You can find a live implementation of this algorithm at *https://gabrielgambetta.com/cgfs/clipping-demo*.

Clipping in the Rendering Pipeline

The order of the chapters in the book is not the order of operations in the rendering pipeline; as explained in the introduction, the chapters are ordered in such a way that visible progress is reached as quickly as possible.

Clipping is a 3D operation; it takes 3D objects in the scene and generates a new set of 3D objects in the scene or, more precisely, it computes the intersection of the scene and the clipping volume. For this reason, clipping must happen after objects have been placed in the scene (that is, using the vertices after the model and camera transforms) but before perspective projection.

The techniques presented in this chapter work reliably, but are very generic. The more prior knowledge you have about your scene, the more efficient your clipping can be. For example, many games pre-process their levels by adding visibility information to them; if you can divide a scene into "rooms," you can make a table listing what rooms are visible from any given room. When rendering the scene later, you just need to figure out what room the camera is in, and you can safely ignore all the rooms marked as "non-visible" from there, saving considerable resources during rendering. The trade-off is, of course, more pre-processing time and a more rigid scene. If you're interested in this topic, read about BSP partitioning and portal systems.

Summary

In this chapter, we finally lifted one of the main limitations caused by the perspective projection equation. We've overcome the limitation that only vertices in front of the camera can be meaningfully projected. In order to do this, we came up with a precise definition of what "being in front of the camera" means: whatever is inside a clipping volume we define with five planes.

Then we developed the equations and algorithms to compute the geometrical intersection between the scene and the clipping volume. As a consequence, we can take an entire scene and remove everything that can't possibly be projected onto the viewport. This not only avoids the cases that can't be handled by the perspective projection equations, it also saves computation resources by removing geometry that would be projected outside of the viewport.

However, after clipping a scene, we might still end up with geometry that *could* be visible in the final canvas, but which *will not*, most likely because there's something else in front of it! We'll find ways to deal with this in the next chapter.

12

HIDDEN SURFACE REMOVAL

We can now render any scene from any point of view, but the resulting image is visually simple: we're rendering objects in wireframe, giving the impression that we're looking at the blueprint of a set of objects, not at the objects themselves.

The remaining chapters of this book focus on improving the visual quality of the rendered scene. By the end of this chapter, we'll be able to render objects that look solid (as opposed to wireframe). We already developed an algorithm to draw filled triangles, but as we will see, using that algorithm correctly in a 3D scene is not as simple as it might seem!

Rendering Solid Objects

The first idea that comes to mind when we want to make solid objects look solid is to use the function `DrawFilledTriangle` that we developed in Chapter 7 to draw each triangle of the objects using a random color (Figure 12-1).

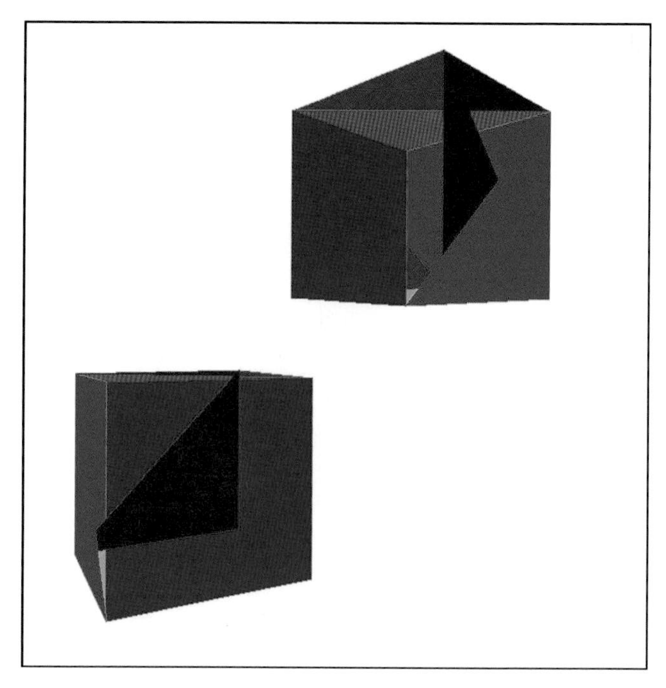

Figure 12-1: Using `DrawFilledTriangle` instead of `DrawWireframeTriangle` doesn't produce the results we expect.

The shapes in Figure 12-1 don't quite look like cubes, do they? If you look closely, you'll see what the problem is: parts of the back faces of the cube are drawn on top of the front faces! This is because we're blindly drawing 2D triangles on the canvas in a "random order" or, more precisely, in the order they happen to be defined in the `Triangles` list of the model, without taking into account the spatial relationships between them.

You might be tempted to go back to the model definition and change the order of the triangles to fix this problem. However, if our scene includes another instance of the cube that is rotated 180°, we'd go back to the original problem. In short, there's no single "correct" triangle order that will work for every instance and camera orientation. What should we do?

Painter's Algorithm

A first solution to this problem is known as the *painter's algorithm*. Real-life painters draw backgrounds first, and then cover parts of them with foreground objects. We could achieve the same effect by drawing all the triangles in the scene back to front. To do so, we'd apply the model and camera transforms and sort the triangles according to their distance to the camera.

This works around the "no single correct order" problem explained above, because now we're looking for a correct ordering for a specific relative position of the objects and the camera.

Although this would indeed draw the triangles in the correct order, it has some drawbacks that make it impractical.

First, it doesn't scale well. The most efficient sorting algorithm known to humans is $O(n \cdot \log(n))$, which means the runtime more than doubles if we double the number of triangles. (To illustrate, sorting 100 triangles would take approximately 200 operations; sorting 200 triangles would take 460, not 400; and sorting 800 triangles would take 2,322 operations, not 1,840!) In other words, this works for small scenes, but it quickly becomes a performance bottleneck as the complexity of the scene increases.

Second, it requires us to know the whole list of triangles at once. This requires a lot of memory and stops us from using a stream-like approach to rendering. We want our renderer to be like a pipeline, where model triangles enter on one end and pixels come out the other end, but this algorithm doesn't start drawing pixels until every triangle has been transformed and sorted.

Third, even if we'd be willing to live with these limitations, there are cases where a correct ordering of triangles just *doesn't exist at all*. Consider the case in Figure 12-2. We will *never* be able to sort these triangles in a way that produces the correct results.

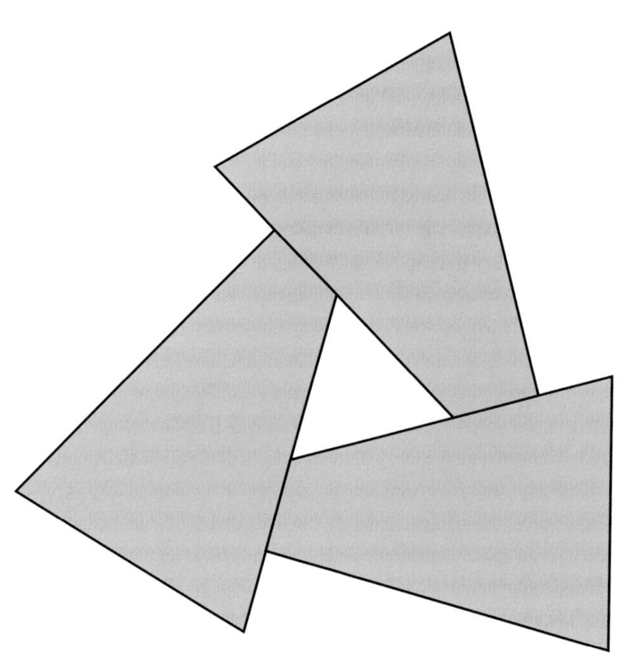

Figure 12-2: There is no way to sort these triangles "back-to-front."

Depth Buffering

We can't solve the ordering problem at the triangle level, so let's try to solve it at the pixel level.

For each pixel on the canvas, we want to paint it with the "correct" color, where the "correct" color is the color of the object that is closest to the camera. In Figure 12-3, that's P_1.

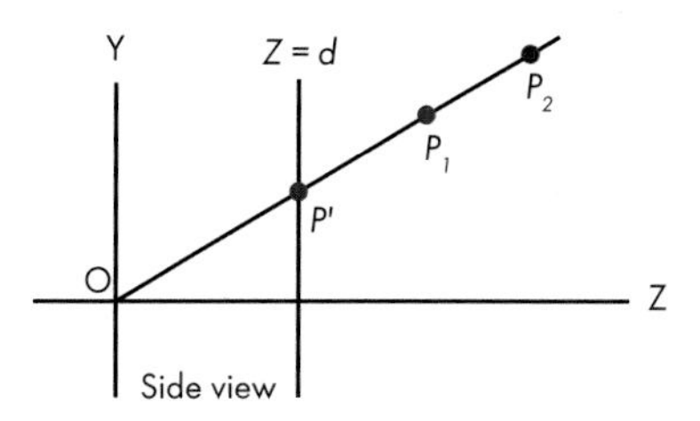

Figure 12-3: Both P_1 *and* P_2 *project to the same* P' *on the canvas. Because* P_1 *is closer to the camera than* P_2*, we want to paint* P' *the color of* P_1*.*

At any time during rendering, each pixel on the canvas represents one point in the scene (before we draw anything, it represents a point infinitely far away). Suppose that for each pixel on the canvas, we kept the Z coordinate of the point it currently represents. When we need to decide whether to paint a pixel with the color of an object, we will do it only if the Z coordinate of the point we're about to paint is smaller than the Z coordinate of the point that is already there. This guarantees that a pixel representing a point in the scene is never drawn over by a pixel representing a point that is farther away from the camera.

Let's go back to Figure 12-3. Suppose that because of the order of the triangles in a model, we want to paint P_2 first and P_1 second. When we paint P_2, the pixel is painted red, and its associated Z value becomes Z_{P_2}. Then we want to paint P_1, and since $Z_{P_2} > Z_{P_1}$, we paint the pixel blue and we get the correct result.

In this particular case, we'd have gotten the correct result regardless of the values of Z, because the points happened to come in a convenient order. But what if we wanted to paint P_1 first and P_2 second? We first paint the pixel blue and store Z_{P_1}; but when we want to paint P_2, we see that $Z_{P_2} > Z_{P_1}$, so we *don't* paint it—because if we did, P_1 would be covered by P_2, which is farther away! We get a blue pixel again, which is the correct result.

In terms of implementation, we need a buffer to store the Z coordinate of every pixel on the canvas; we call this the *depth buffer*. It has the same dimensions as the canvas, but its elements are real numbers representing depth values, not pixels.

But where do the Z values come from? These should be the Z values of the points after they're transformed but before they're perspective-projected. However, this only gives us Z values for vertices; we need a Z value for every pixel of every triangle.

Here is yet another application of the attribute-mapping algorithm we developed in Chapter 8. Why not use *Z* as the attribute and interpolate it across the face of the triangle, just like we did before with color-intensity values? By now you know how to do it: take the values of `z0`, `z1`, and `z2`; compute `z01`, `z02`, and `z012`; combine them to get `z_left` and `z_right`; then, for each horizontal segment, compute `z_segment`. Finally, instead of blindly calling `PutPixel(x, y, color)`, we do this:

```
z = z_segment[x - xl]
if (z < depth_buffer[x][y]) {
    canvas.PutPixel(x, y, color)
    depth_buffer[x][y] = z
}
```

For this to work correctly, every entry in `depth_buffer` should be initialized to $+\infty$ (or just "a very big value"). This guarantees that the first time we want to draw a pixel, the condition will be true, because any point in the scene is closer to the camera than a point infinitely far away.

The results we get now are much better—check out Figure 12-4.

Figure 12-4: The cubes now look like cubes, regardless of the ordering of their triangles.

You can find a live implementation of this algorithm at *https://gabrielgambetta.com/cgfs/depth-demo*.

Using 1/Z instead of Z

The results look much better, but what we're doing is subtly wrong. The values of Z for the vertices are correct (they come from data, after all), but in most cases the linearly interpolated values of Z for the rest of the pixels are incorrect. This might not even result in a visible difference at this point, but it would become an issue later.

To see how the values are wrong, consider the simple case of a line segment from $A(-1, 0, 2)$ to $B(1, 0, 10)$, with its midpoint M at $(0, 0, 6)$. Specifically, because M is the midpoint of AB, we know that $M_z = (A_z + B_z)/2 = 6$. Figure 12-5 shows this line segment.

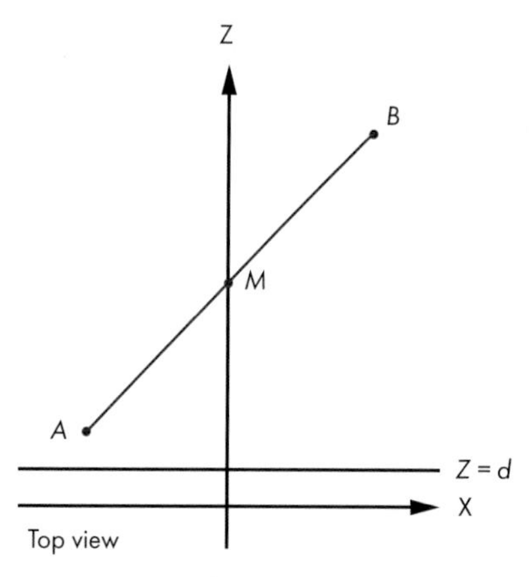

Figure 12-5: A line segment AB *and its midpoint* M

Let's compute the projection of these points with $d = 1$. Applying the perspective projection equations, we get $A'_x = A_x/A_z = -1/2 = -0.5$. Similarly, $B'_x = 0.1$ and $M'_x = 0$. Figure 12-6 shows the projected points.

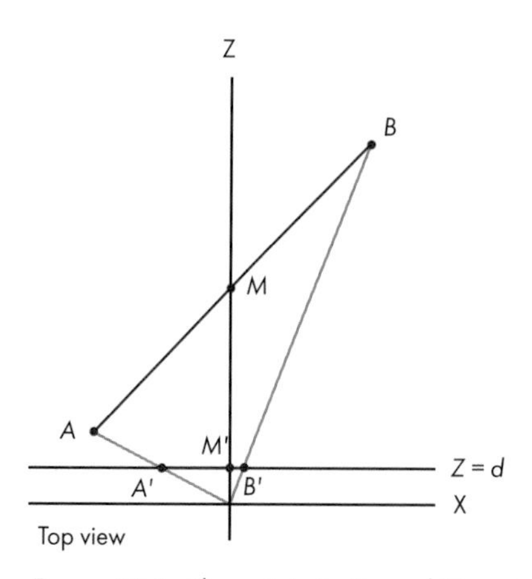

Figure 12-6: The points A, B, *and* M *projected onto the projection plane*

$A'B'$ is a horizontal segment on the viewport. We know the values of A_z and B_z. Let's see what happens if we try to compute the value of M_z using linear interpolation. The implied linear function looks like Figure 12-7.

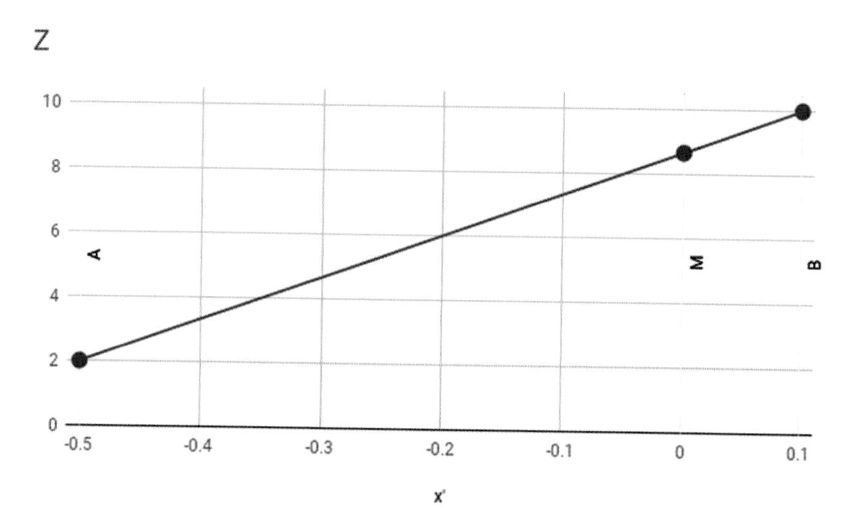

Figure 12-7: The values of A_z and B_z for $A_{x'}$ and $B_{x'}$ define a linear function z = f(x').

The slope of the function is constant, so we can write

$$\frac{M_z - A_z}{M'_x - A'_x} = \frac{B_z - A_z}{B'_x - A'_x}$$

We can manipulate that expression to solve for M_z:

$$M_z = A_z + (M'_x - A'_x)\left(\frac{B_z - A_z}{B'_x - A'_x}\right)$$

If we plug in the values we know and do some arithmetic, we get

$$M_z = 2 + (0 - (-0.5))\left(\frac{10 - 2}{0.1 - (-0.5)}\right) = 2 + (0.5)\left(\frac{8}{0.6}\right) = 8.666$$

This says that the value of M_z is 8.666, but we know for a fact it's actually 6!

Where did we go wrong? We're using linear interpolation, which we know works well, and we're feeding it the correct values, which come from data, so why is the result wrong?

Our mistake is hidden in the implicit assumption we make when we use linear interpolation: that the function we are interpolating is linear to begin with! In this case, it turns out it isn't.

If $Z = f(x', y')$ was a linear function of x' and y', we could write it as $Z = Ax' + By' + C$ for some values of A, B, and C. This kind of function has the property that the difference of its value between two points depends on the difference between the points but not on the points themselves:

$$f(x' + \Delta x, y' + \Delta y) - f(x', y') = [A(x' + \Delta x) + B(y' + \Delta y) + C] - [A \cdot x' + B \cdot y' + C]$$

$$= A(x' + \Delta x - x') + B(y' + \Delta y - y') + C - C$$

$$= A\Delta x + B\Delta y$$

That is, for a given difference in screen coordinates, the difference in Z would always be the same.

More formally, the equation of the plane that contains the line segment we're studying is

$$Ax + By + Cz + D = 0$$

On the other hand we have the perspective projection equations:

$$x' = \frac{x \cdot d}{z}$$

$$y' = \frac{x \cdot d}{z}$$

We can get x and y back from these:

$$x = \frac{z \cdot x'}{d}$$

$$y = \frac{z \cdot y'}{d}$$

If we replace x and y in the plane equation with these expressions, we get

$$\frac{Ax'z + By'z}{d} + Cz + D = 0$$

Multiplying by d and then solving for z,

$$Ax'z + By'z + dCz + dD = 0$$

$$(Ax' + By' + dC)z + dD = 0$$

$$z = \frac{-dD}{Ax' + By' + dC}$$

This is clearly *not* a linear function of x' and y', and this is why linearly interpolating values of z gave us an incorrect result.

However, if we compute $1/z$ instead of z, we get

$$1/z = \frac{Ax' + By' + dC}{-dD}$$

This clearly *is* a linear function of x' and y'. This means we could linearly interpolate values of $1/z$ and get the correct results.

In order to verify that this works, let's calculate the interpolated value for M_z, but this time using the linear interpolation of $1/z$:

$$\frac{M_{\frac{1}{z}} - A_{\frac{1}{z}}}{M'_x - A'_x} = \frac{B_{\frac{1}{z}} - A_{\frac{1}{z}}}{B'_x - A'_x}$$

$$M_{\frac{1}{z}} = A_{\frac{1}{z}} + (M'_x - A'_x)(\frac{B_{\frac{1}{z}} - A_{\frac{1}{z}}}{B'_x - A'_x})$$

$$M_{\frac{1}{z}} = \frac{1}{2} + (0 - (-0.5))(\frac{\frac{1}{10} - \frac{1}{2}}{0.1 - (-0.5)}) = 0.166666$$

And therefore

$$M_z = \frac{1}{M_{\frac{1}{z}}} = \frac{1}{0.166666} = 6$$

This value is correct, in the sense that it matches our original calculation of M_z based on the geometry of the line segment.

All of this means we need to use values of $1/z$ instead of values of z for depth buffering. The only practical differences in the pseudocode are that every entry in the buffer should be initialized to 0 (which is conceptually $1/+\infty$), and that the comparison should be inverted (we keep the bigger value of $1/z$, which corresponds to a smaller value of z).

Back Face Culling

Depth buffering produces the desired results. But can we make things even faster?

Going back to the cube, even if each pixel ends up having the right color, many of them are painted over several times. For example, if a back face of the cube is rendered before a front face, many pixels will be painted twice. This can be costly. So far we've been computing $1/z$ for every pixel, but soon we'll add more attributes, such as illumination. As the number of per-pixel operations we need to perform increases, computing pixels that will never be visible becomes more and more wasteful.

Can we discard pixels earlier, before we go into all of this computation? It turns out we can discard *entire triangles* before we even start rendering!

So far we've been talking informally about *front faces* and *back faces*. Imagine every triangle has two distinct sides; it's impossible to see both sides of a triangle at the same time. In order to distinguish between the two sides, we'll stick an imaginary arrow on each triangle, perpendicular to its surface. Then we'll take the cube and make sure every arrow is pointing out. Figure 12-8 shows this idea.

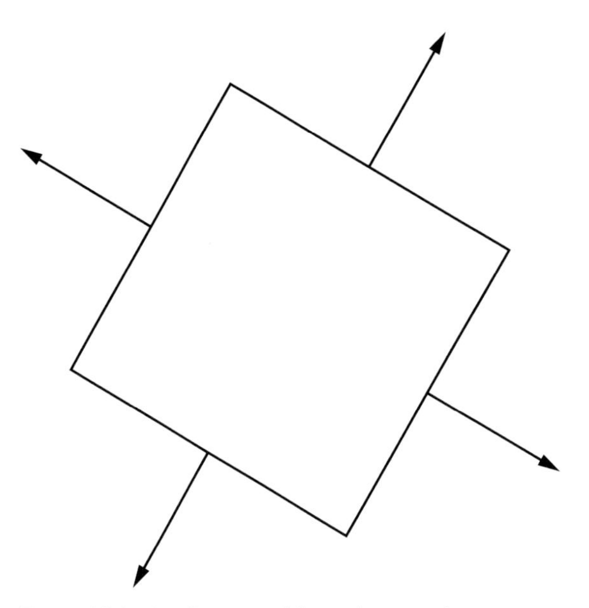

Figure 12-8: A cube viewed from above, with arrows on each triangle pointing out

These arrows let us classify each triangle as "front" or "back," depending on whether they point toward the camera or away from the camera. More formally, if the view vector and this arrow (which is actually a normal vector of the triangle) form an angle of less than 90°, the triangle is front-facing; otherwise, it's back-facing (Figure 12-9).

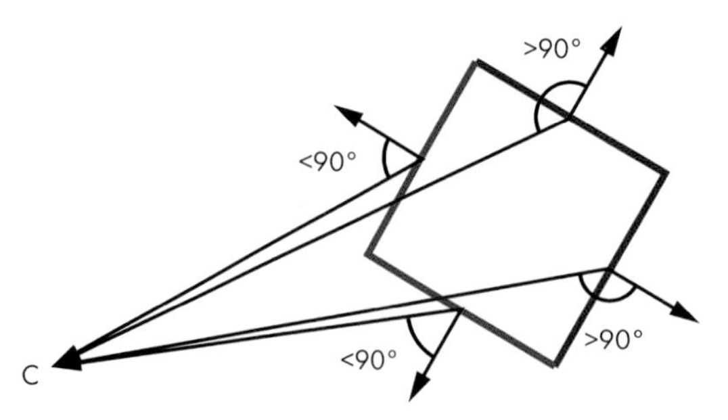

Figure 12-9: The angle between the view vector and the normal vector of a triangle lets us classify it as front-facing or back-facing.

At this point, we need to impose a restriction on our 3D models: that they are *closed*. The exact definition of closed is pretty involved, but fortunately an intuitive understanding is enough. The cube we've been working with is closed; we can only see its exterior. If we removed one of its faces, it wouldn't be closed because we could see inside it. This doesn't mean we can't have objects with holes or concavities; we would just model these with thin "walls." See Figure 12-10 for some examples.

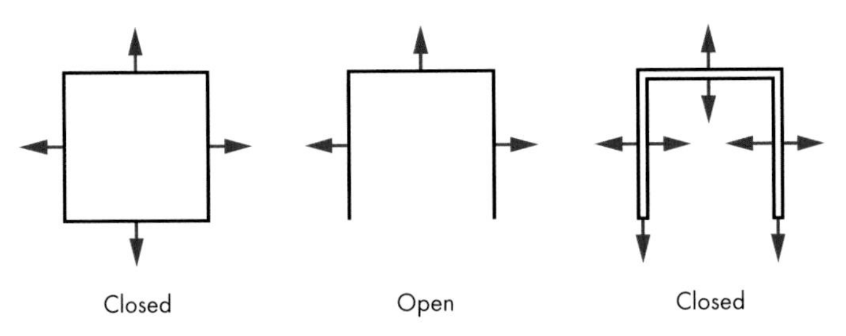

Figure 12-10: Some examples of open and closed objects

Why impose this restriction? Closed objects have the interesting property that the set of front faces completely covers the set of back faces, no matter the orientation of the model or the camera. This means we don't need to draw the back faces at all, saving valuable computation time.

Since we can discard (cull) all the back faces, this algorithm is called *back face culling*. Its pseudocode is remarkably simple for an algorithm that can cut our rendering time by half!

```
CullBackFaces(object, camera) {
  for T in object.triangles {
    if T is back-facing {
      remove T from object.triangles
    }
  }
}
```

Listing 12-1: The back face culling algorithm

Let's take a more detailed look at how to determine whether a triangle is front-facing or back-facing.

Classifying Triangles

Suppose we have the normal vector $\vec{N}$ of a triangle and the vector $\vec{V}$ from a vertex of the triangle to the camera. Now suppose $\vec{N}$ points to the outside of the object. In order to classify the triangle as front-facing or back-facing, we compute the angle between $\vec{N}$ and $\vec{V}$ and check whether they're within 90° of each other.

We can again use the properties of the dot product to make this simpler. Remember that if α is the angle between $\vec{N}$ and $\vec{V}$, then

$$\frac{\langle \vec{N}, \vec{V} \rangle}{|\vec{N}|\,|\vec{V}|} = \cos(\alpha)$$

Because $\cos(\alpha)$ is non-negative for $|\alpha| \leq 90°$, we only need to know the sign of this expression to classify a triangle as front-facing or back-facing. Note that $|\vec{N}|$ and $|\vec{V}|$ are always positive, so they don't affect the sign of the expression. Therefore

$$\text{sign}(\langle \vec{N}, \vec{V} \rangle) = \text{sign}(\cos(\alpha))$$

The classification criterion is simply this:

$\langle \vec{N}, \vec{V} \rangle \leq 0$	Back-facing
$\langle \vec{N}, \vec{V} \rangle > 0$	Front-facing

The edge case $\langle \vec{N}, \vec{V} \rangle = 0$ corresponds to the case where we're looking at the edge of a triangle head on—that is, when the camera and the triangle are coplanar. We can classify this triangle either way without affecting the result much, so we choose to classify it as back-facing to avoid dealing with degenerate triangles.

Where do we get the normal vector from? It turns out there's a vector operation, the *cross product* $\vec{A} \times \vec{B}$, that takes two vectors $\vec{A}$ and $\vec{B}$ and produces a vector perpendicular to both (for a definition of this operation, see the Linear Algebra appendix. In other words, the cross product of two vectors on the surface of a triangle is a normal vector of that triangle. We can easily get two vectors on the triangle by subtracting its vertices from each other. So computing the direction of the normal vector of the triangle ABC is straightforward:

$$\vec{V_1} = B - A$$

$$\vec{V_2} = C - A$$

$$\vec{N} = \vec{V_1} \times \vec{V_2}$$

Note that "the direction of the normal vector" is not the same as "the normal vector." There are two reasons for this. The first one is that $|\vec{N}|$ isn't necessarily equal to 1. This isn't really important because normalizing $\vec{N}$ would be trivial and because we only care about the sign of $\langle \vec{N}, \vec{V} \rangle$.

The second reason is that if $\vec{N}$ is a normal vector of ABC, so is $-\vec{N}$, and in this case we care deeply about the direction $\vec{N}$ points in, because this is exactly what lets us classify triangles as either front-facing or back-facing.

Moreover, the cross product of two vectors is not commutative: $\vec{V_1} \times \vec{V_2} = -(\vec{V_2} \times \vec{V_1})$. In other words, the order of the vectors in this operation matters. And since we defined V_1 and V_2 in terms of A, B, and C, this means the order of the vertices in a triangle matters. We can't treat the triangles ABC and ACB as the same triangle anymore.

Fortunately, none of this is random. Given the definition of the cross product operation, the way we defined V_1 and V_2, and the coordinate system we use (X to the right, Y up, Z forward), there is a very simple rule that determines the direction of the normal vector: if the vertices of the triangle ABC are in clockwise order when you look at them from the camera, the normal vector as calculated above will point toward the camera—that is, the camera is looking at the front face of the triangle.

We just need to keep this rule in mind when designing 3D models manually and list the vertices of each triangle in clockwise order when looking at its front face, so that their normals point "out" when we compute them this way. Of course, the example cube model we've been using so far follows this rule.

Summary

In this chapter, we made our renderer, which could previously only render wireframe objects, capable of rendering solid-looking objects. This is more involved than just using `DrawFilledTriangle` instead of `DrawWireframeTriangle`, because we need triangles close to the camera to obscure triangles further away from the camera.

The first idea we explored was to draw the triangles from back to front, but this had a few drawbacks that we discussed. A better idea is to work at the pixel level; this idea led us to a technique called depth buffering, which produces correct results regardless of the order in which we draw the triangles.

We finally explored an optional but valuable technique that doesn't change the correctness of the results, but can save us from rendering approximately half of the triangles of the scene: back face culling. Since all the back-facing triangles of a closed object are covered by all its front-facing triangles, there's no need to draw the back-facing triangles at all. We presented a simple algebraic way to determine whether a triangle is front- or back-facing.

Now that we can render solid-looking objects, we'll devote the rest of this book to making these objects look more realistic.

13

SHADING

Let's continue making our images more realistic; in this chapter, we'll examine how to add lights to the scene and how to illuminate the objects it contains. First, let's look at a bit of terminology.

Shading vs. Illumination

The title of this chapter is "Shading," not "Illumination"; these are two different but closely related concepts. *Illumination* refers to the math and algorithms necessary to compute the effect of light on a single point in the scene; *shading* deals with techniques that extend the effect of light on a discrete set of points to entire objects.

In Chapter 3, we looked at all we need to know about illumination. We can define ambient, point, and directional lights, and we can compute the illumination at any point in the scene given its position and a surface normal at that point:

$$I_P = I_A + \sum_{i=1}^{n} I_i \cdot \left[\frac{\langle \vec{N}, \vec{L_i} \rangle}{|\vec{N}||\vec{L_i}|} + \left(\frac{\langle \vec{R_i}, \vec{V} \rangle}{|\vec{R_i}||\vec{V}|} \right)^s \right]$$

This illumination equation expresses how light illuminates a point in the scene. The way this worked in our raytracer is exactly the same way it works in our rasterizer.

The more interesting part, which we'll explore in this chapter, is how to extend the "illumination at a point" algorithms we developed into "illumination at every point of a triangle" algorithms.

Flat Shading

Let's start simple. Since we can compute illumination at a point, we can just pick any point in a triangle (for example, its center), compute the illumination at that point, and use it to shade the whole triangle. To do the actual shading, we can multiply the color of the triangle by the illumination value. Figure 13-1 shows the results.

Figure 13-1: In flat shading, we compute illumination at the center of the triangle and use it for the entire triangle.

The results are promising. Every point in a triangle has the same normal, so as long as a light is reasonably far from it, the light vectors for every point are *approximately* parallel and every point receives *approximately* the same amount of light. The discontinuity between the two triangles that make up each side of the cube, especially visible on the green face in Figure 13-1, is a consequence of the light vectors being *approximately*, but not *exactly*, parallel.

So what happens if we try this technique with an object for which every point has a different normal, like the sphere in Figure 13-2?

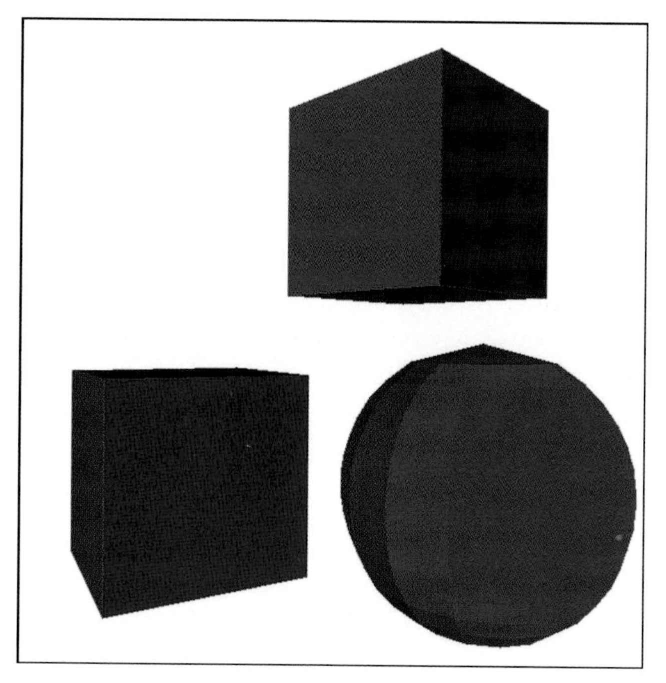

Figure 13-2: Flat shading works reasonably well for objects with flat faces, but not so well for objects that are supposed to be curved.

Not so good. It is very obvious that the object is not a true sphere, but an approximation made out of flat, triangular patches. Because this kind of illumination makes curved objects look flat, it's called *flat shading*.

Gouraud Shading

How can we remove these discontinuities in lighting? Instead of computing illumination only at the center of a triangle, we can compute illumination at its three vertices. This gives us three illumination values between 0.0 and 1.0, one for each vertex of the triangle. This puts us in exactly the same situation as Chapter 8: we can use `DrawShadedTriangle` directly, using the illumination values as the "intensity" attribute.

This technique is called *Gouraud shading*, after Henri Gouraud, who came up with the idea in 1971. Figure 13-3 shows the results of applying it to the cube and the sphere.

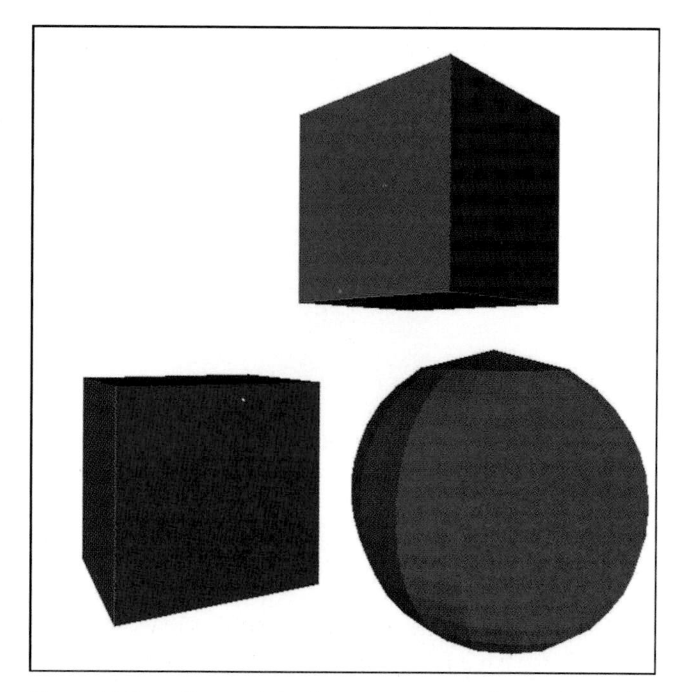

Figure 13-3: In Gouraud shading, we compute illumination at the vertices of the triangle and interpolate them across its surface.

The cube looks better: the discontinuity is gone, because both triangles of each face share two vertices and they have the same normal, so the illumination at these two vertices is identical for both triangles.

The sphere, however, still looks faceted, and the discontinuities on its surface look really wrong. This shouldn't be surprising: we're treating the sphere as a collection of flat surfaces. In particular, despite every triangle sharing vertices with its neighboring triangles, they have different normals. Figure 13-4 shows the problem.

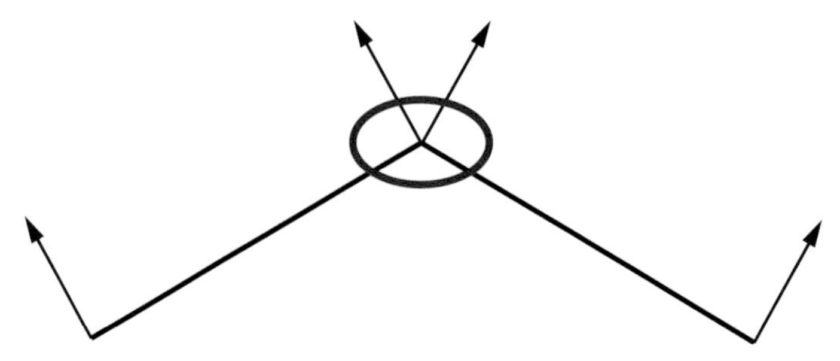

Figure 13-4: We get two different values for the illumination at the shared vertex, because they depend on the normals of the triangles, which are different.

Let's take a step back. The fact that we're using flat triangles to represent a curved object is a limitation of our techniques, not a property of the object itself.

Each vertex in the sphere model corresponds to a point on the sphere, but the triangles they define are just an approximation of its surface. It would be a good idea to make the vertices in the model represent the points in the sphere as closely as possible. That means, among other things, using the actual sphere normals for each vertex, as shown in Figure 13-5.

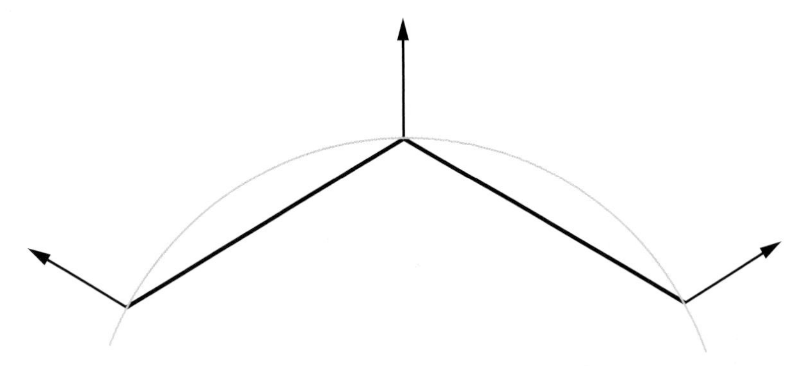

Figure 13-5: We can give each vertex the normal of the curved surface it represents.

Note that this doesn't apply to the cube; even though triangles share vertex positions, each face needs to be shaded independently of the others. There's no single "correct" normal for the vertices of a cube.

Our renderer has no way to know whether a model is supposed to be an approximation of a curved object or the exact representation of a flat one. After all, a cube is a very crude approximation of a sphere! To solve this, we'll make the triangle normals part of the model, so its designer can make this decision.

Some objects, like the sphere, have a single normal per vertex. Other objects, like the cube, have a different normal for each triangle that uses the vertex. So we can't make the normals a property of the vertices; they need to be a property of the triangles that use them:

```
model {
    name = cube
    vertices {
        0 = (-1, -1, -1)
        1 = (-1, -1,  1)
        2 = (-1,  1,  1)
        ...
    }
    triangles {
        0 = {
```

```
            vertices = [0, 1, 2]
            normals = [(-1, 0, 0), (-1, 0, 0), (-1, 0, 0)]
        }
        ...
    }
}
```

Figure 13-6 shows the scene rendered using Gouraud shading and the appropriate vertex normals.

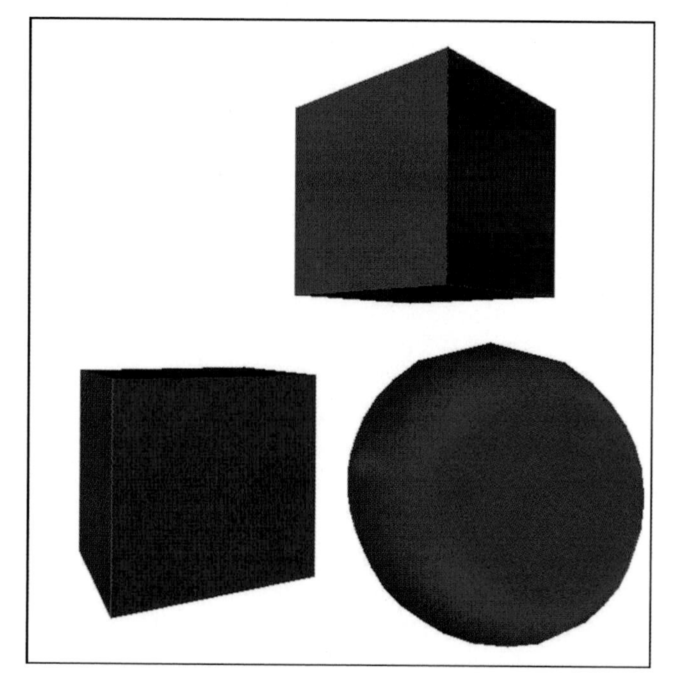

Figure 13-6: Gouraud shading with normal vectors specified in the model. The cubes still look like cubes, and the sphere now looks like a sphere.

The cubes still look like cubes, and the sphere now looks remarkably like a sphere. In fact, you can only tell it's made out of triangles by looking at its outline. This could be improved by using more, smaller triangles, at the expense of requiring more computing power.

Gouraud shading starts breaking down when we try to render shiny objects, though; the specular highlight on the sphere is decidedly unrealistic.

This is an indication of a more general problem. When we move a point light very close to a big face, we'd naturally expect it to look brighter and the specular effects to become more pronounced; however, Gouraud shading produces the exact opposite (Figure 13-7).

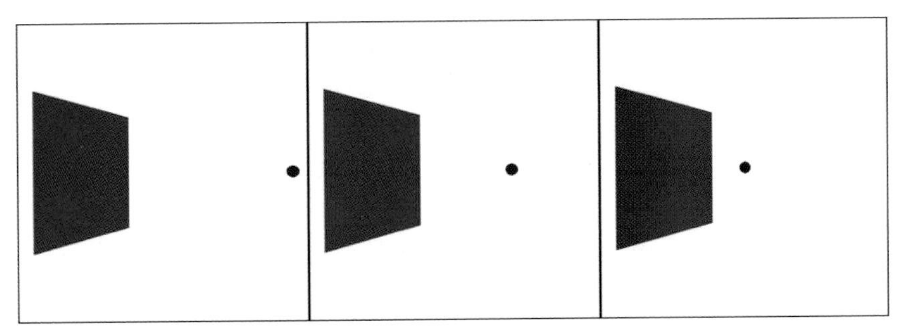

Figure 13-7: Contrary to our expectations, the closer the point light is to a face, the darker it looks.

We expect points near the center of the triangle to receive a lot of light, because $\vec{L}$ and $\vec{N}$ are roughly parallel. However, we're not computing lighting at the center of the triangle, but at its vertices. There, the closer the light is to the surface, the *bigger* the angle with the normal, so they receive little illumination. This means that every interior pixel will end up with an intensity value that is the result of interpolating between two small values, which is also a low value, as shown in Figure 13-8.

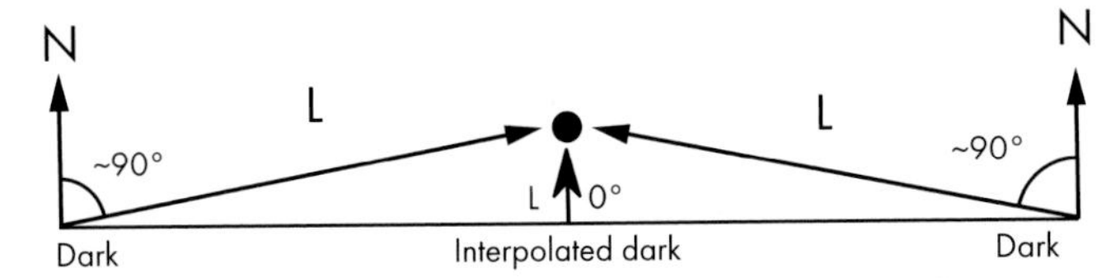

Figure 13-8: Interpolating illumination from the vertices, which are dark, results in a dark center, although the normal is parallel to the light vector at that point.

So, what to do?

Phong Shading

We can overcome the limitations of Gouraud shading, but as usual, there's a trade-off between quality and resource usage.

Flat shading involved a single illumination calculation per triangle. Gouraud shading requires three illumination calculations per triangle, plus the interpolation of a single attribute, the illumination, across the triangle. The next step in quality requires us to calculate illumination at every pixel of the triangle.

This doesn't sound particularly complex from a theoretical point of view; we're computing lighting at one or three points already, and we were computing per-pixel lighting for the raytracer after all. What's tricky here is figuring out where the inputs to the illumination equation come from.

Recall that the full illumination equation, with ambient, diffuse, and specular components, is:

$$I_P = I_A + \sum_{i=1}^{n} I_i \left(\frac{\langle \vec{N}, \vec{L_i} \rangle}{|\vec{N}|\,|\vec{L_i}|} + \left(\frac{\langle \vec{R}, \vec{V} \rangle}{|\vec{R}|\,|\vec{V}|} \right)^s \right)$$

First, we need $\vec{L}$. For directional lights, $\vec{L}$ is given. For point lights, $\vec{L}$ is defined as the vector from the point in the scene, P, to the position of the light, Q. However, we don't have Q for every pixel of the triangle, but only for the vertices.

What we do have is the projection of P; that is, the x' and y' we're about to draw on the canvas! We know that

$$x' = \frac{xd}{z}$$

$$y' = \frac{xd}{z}$$

We also happen to have an interpolated but geometrically correct value for $\frac{1}{z}$ as part of the depth-buffering algorithm, so

$$x' = xd\frac{1}{z}$$

$$y' = yd\frac{1}{z}$$

We can recover P from these values:

$$x = \frac{x'}{d\frac{1}{z}}$$

$$y = \frac{y'}{d\frac{1}{z}}$$

$$z = \frac{1}{\frac{1}{z}}$$

We also need $\vec{V}$. This is the vector from the camera (which we know) to P (which we just computed), so $\vec{V}$ is simply $P - C$.

Next, we need $\vec{N}$. We only know the normals at the vertices of the triangle. When all you have is a hammer, every problem looks like a nail; our hammer is—you probably guessed it—linear interpolation of attribute values. So let's take the values of N_x, N_y, and N_z at each vertex and treat each of them as an attribute we can linearly interpolate. Then, at every pixel, we reassemble the interpolated components into a vector, normalize it, and use it as the normal at that pixel.

This technique is called *Phong shading*, after Bui Tuong Phong, who invented it in 1973. Figure 13-9 shows the results.

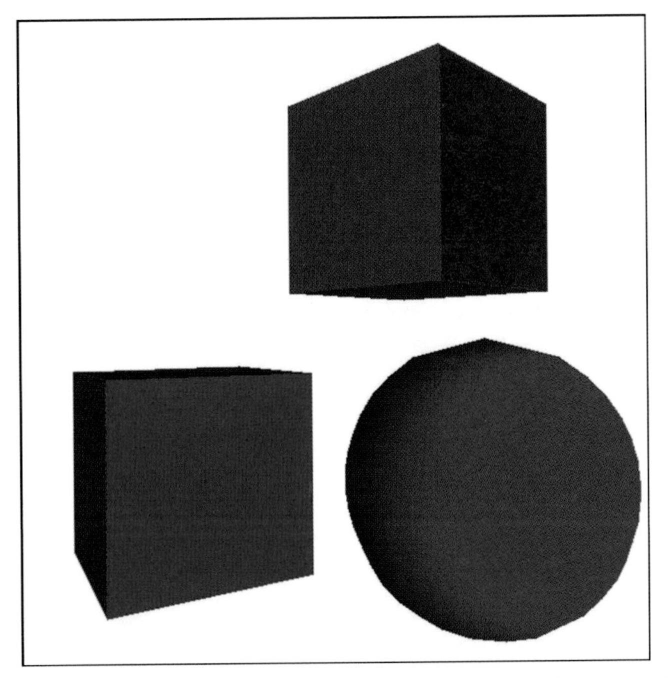

Figure 13-9: Phong shading. The surface of the sphere looks smooth and the specular highlight is clearly visible.

You can find a live implementation of this algorithm at *https://gabrielgambetta.com/cgfs/shading-demo*.

The sphere looks much better now. Its surface displays the proper curvature, and the specular highlights look well defined. The contour, however, still betrays the fact that we're rendering an approximation made of triangles. This is not a shortcoming of the shading algorithm, which only determines the color of each pixel of the surface of the triangles but has no control over the shape of the triangles themselves. This sphere approximation uses 420 triangles; we could get a smoother contour by using more triangles, at the cost of worse performance.

Phong shading also solves the problem with the light getting close to a face, now giving the expected results (Figure 13-10).

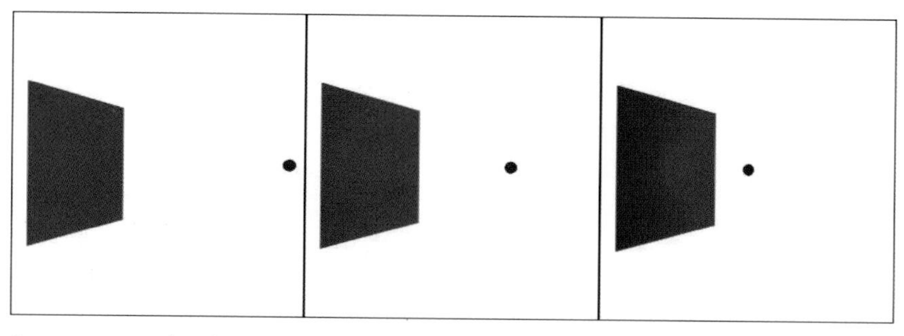

Figure 13-10: The closer the light is to the surface, the brighter and better defined the specular highlight looks.

At this point, we've matched the capabilities of the raytracer developed in Part I, except for shadows and reflections. Using the exact same scene definition, Figure 13-11 shows the output of the rasterizer we're developing.

Figure 13-11: The reference scene, rendered by the rasterizer

For reference, Figure 13-12 shows the raytraced version of the same scene.

Figure 13-12: The reference scene, rendered by the raytracer

The two versions look almost identical, despite using vastly different techniques. This is expected, since the scene definition is identical. The only visible difference can be found in the contour of the spheres: the raytracer renders them as mathematically perfect objects, but we use an approximation made of triangles for the rasterizer.

Another difference is the performance of the two renderers. This is very hardware- and implementation-dependent, but generally speaking, rasterizers can produce full-screen images of complex scenes up to 60 times per second or more, which makes them suitable for interactive applications such as videogames, while raytracers may take multiple seconds to render the same scene once. This difference might tend to disappear in the future; advances in hardware in recent years are making raytracer performance much more competitive with that of rasterizers.

Summary

In this chapter, we added illumination to our rasterizer. The illumination equation we use is exactly the same as the one in Chapter 3, because we're using the same lighting model. However, where the raytracer computed the illumination equation at each pixel, our rasterizer can support a variety of different techniques to achieve a specific trade-off between performance and image quality.

The fastest shading algorithm, which also produces the least appealing results, is flat shading: we compute the illumination of a single point in a triangle and use it for every pixel in that triangle. This results in a very faceted appearance, especially for objects that approximate curved surfaces such as spheres.

One step up the quality ladder, we have Gouraud shading: we compute the illumination of the three vertices of a triangle and then interpolate this value across the face of the triangle. This gives objects a smoother appearance, including curved objects. However, this technique fails to capture more subtle lighting effects, such as specular highlights.

Finally, we studied Phong shading. Much like our raytracer, it computes the illumination equation at every pixel, producing the best results and also the worst performance. The trick in Phong shading is knowing how to compute all the necessary values to evaluate the illumination equation; once again, the answer is linear interpolation—in this case, of normal vectors.

In the next chapter, we'll add even more detail to the surface of our triangles, using a technique that we haven't studied for the raytracer: texture mapping.

14

TEXTURES

Our rasterizer can render objects like cubes or spheres. But we usually don't want to render abstract geometric objects like cubes and spheres; instead, we want to render real-world objects, like crates and planets or dice and marbles. In this chapter, we'll look at how we can add visual detail to the surface of our objects by using *textures*.

Painting a Crate

Let's say we want our scene to have a wooden crate. How do we turn a cube into a wooden crate? One option is to add a lot of triangles to replicate the grain of the wood, the heads of the nails, and so on. This would work, but it would add a lot of geometric complexity to the scene, resulting in a big performance hit.

Another option is to fake the details: instead of modifying the geometry of an object, we just "paint" something that looks like wood on top of it. Unless you're looking at the crate from up close, you won't notice the difference, and the computational cost is significantly lower than adding lots of geometric detail.

Note that the two options aren't incompatible: you can choose the right balance between adding geometry and painting on that geometry to achieve the image quality and performance you require. Since we know how to deal with geometry, we'll explore the second option.

First, we need an image to paint on our triangles; in this context, we call this image a *texture*. Figure 14-1 shows a wooden crate texture.

Figure 14-1: Wooden crate texture (by Filter Forge—Attribution 2.0 Generic (CC BY 2.0) license)

Next, we need to specify how this texture is applied to the model. We can define this mapping on a per-triangle basis, by specifying which points of the texture should go on each vertex of the triangle (Figure 14-2).

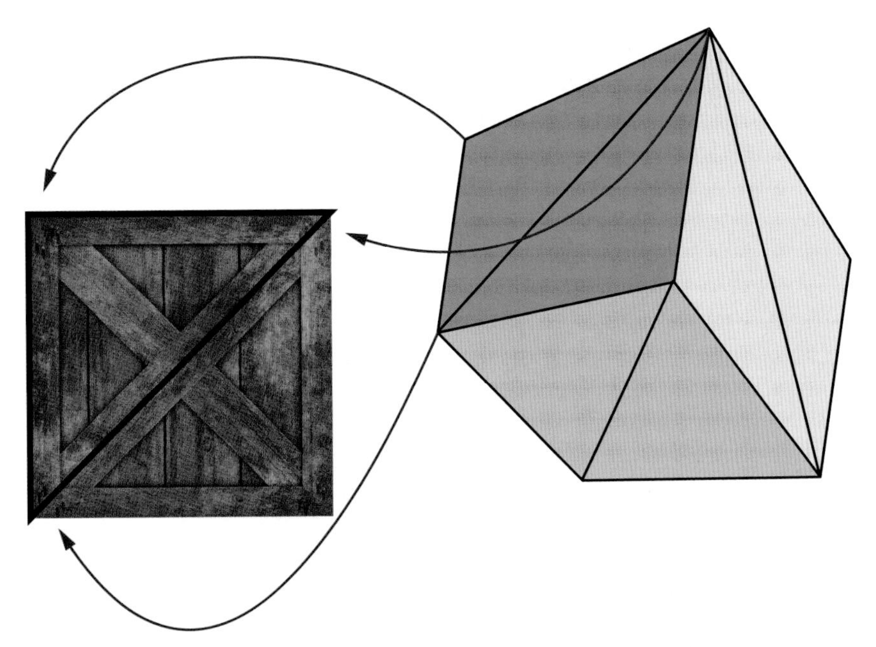

Figure 14-2: We associate a point in the texture with each vertex of the triangle.

To define this mapping, we need a coordinate system to refer to points in the texture. Remember, a texture is just an image, represented as a rectangular array of pixels. We could use x and y coordinates and talk about pixels in the texture, but we're already using these names for the canvas. Therefore, we use u and v for the texture coordinates and we call the texture's pixels *texels* (a contraction of *tex*ture *el*ements).

We'll fix the origin of this (u, v) coordinate system at the top-left corner of the texture. We'll also declare that u and v are real numbers in the range $[0, 1]$, regardless of the actual texel dimensions of the texture. This is very convenient for several reasons. For example, we may want to use a lower- or higher-resolution texture depending on how much RAM we have available; because we're not tied to the actual pixel dimensions, we can change resolutions without having to modify the model itself. We can multiply u and v by the texture width and height respectively to get the actual texel indices tx and ty.

The basic idea of texture mapping is simple: we compute the (u, v) coordinates for each pixel of the triangle, fetch the appropriate texel from the texture, and paint the pixel with that color. But the model only specifies u and v coordinates for the three vertices of the triangle, and we need them for each pixel . . .

By now you can probably see where this is going. Yes, it's our good friend linear interpolation. We can use attribute mapping to interpolate the values of u and v across the face of the triangle, giving us (u, v) at each pixel. From this we can compute (tx, ty), fetch the texel, apply shading, and paint the pixel with the resulting color. You can see the result of doing this in Figure 14-3.

Figure 14-3: The texture looks deformed when applied to the objects.

The results are a little underwhelming. The exterior shape of the crates looks fine, but if you pay close attention to the diagonal planks, you'll notice they look deformed, as if bent in weird ways. What went wrong?

As in Chapter 12, we made an implicit assumption that turns out not to be true: namely, that u and v vary linearly across the screen. This is clearly not the case. Consider the wall of a very long corridor painted with alternating vertical black and white stripes. As the wall recedes into the distance, the vertical stripes should look thinner and thinner. If we make the u coordinate vary linearly with x', we get incorrect results, as illustrated in Figure 14-4.

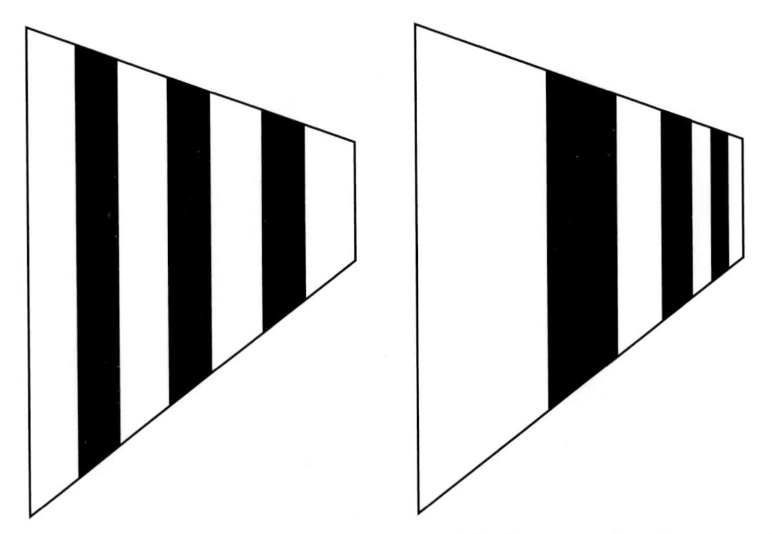

Figure 14-4: Linear interpolation of u and v (left) doesn't produce the expected perspective-correct results (right).

The situation is very similar to the one we encountered in Chapter 12, and the solution is also very similar: although u and v aren't linear in screen coordinates, $\frac{u}{z}$ and $\frac{v}{z}$ are. (The proof is very similar to the $\frac{1}{z}$ proof: consider that u varies linearly in 3D space, and substitute x and y with their screen-space expressions.) Since we already have interpolated values of $\frac{1}{z}$ at each pixel, it's enough to interpolate $\frac{u}{z}$ and $\frac{v}{z}$ and get u and v back:

$$u = \frac{\frac{u}{z}}{\frac{1}{z}}$$

$$v = \frac{\frac{v}{z}}{\frac{1}{z}}$$

This produces the result we expect, as you can see in Figure 14-5.

Figure 14-5: Linear interpolation of u/z *and* v/z *does produce perspective-correct results.*

Figure 14-6 shows the two results side by side, to make it easier to appreciate the difference.

Figure 14-6: A comparison of the "linear u *and* v*" result (left) and the "linear* u/z *and* v/z*" result (right)*

You can find a live implementation of this algorithm at *https://gabrielgambetta.com/cgfs/textures-demo*.

These examples look nice because the size of the texture and the size of the triangles we're applying it to, measured in pixels, is roughly similar. But what happens if the triangle is several times bigger or smaller than the texture? We'll explore those situations next.

Bilinear Filtering

Suppose we place the camera very close to one of the cubes. We'll see something like Figure 14-7.

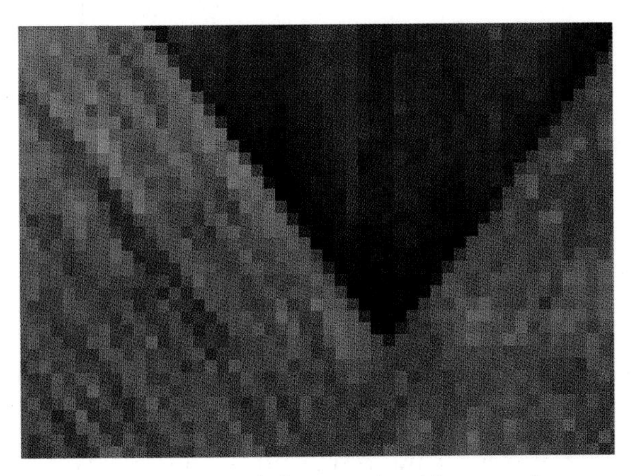

Figure 14-7: A textured object rendered from up close

The image looks very blocky. Why does this happen? The triangle on the screen has more pixels than the texture has texels, so each texel is mapped to many consecutive pixels.

We are interpolating texture coordinates u and v, which are real values between 0.0 and 1.0. Later, given the texture dimensions w and h, we map the u and v coordinates to tx and ty texel coordinates by multiplying them by w and h respectively. But because a texture is an array of pixels with integer indices, we round tx and ty down to the nearest integer. For this reason, this basic technique is called *nearest neighbor filtering*.

Even if (u, v) varies smoothly across the face of the triangle, the resulting texel coordinates "jump" from one whole pixel to the next, causing the blocky appearance we can see in Figure 14-7.

We can do better. Instead of rounding tx and ty down, we can interpret a fractional texel coordinate (tx, ty) as describing a position *between* four integer texel coordinates (obtained by the combinations of rounding tx and ty up and down). We can take the four colors of the surrounding integer texels, and compute a linearly interpolated color for the fractional texel. This will produce a noticeably smoother result (Figure 14-8).

Figure 14-8: A textured object rendered from up close, using interpolated colors

Let's call the four surrounding pixels *TL*, *TR*, *BL*, and *BR* (for top-left, top-right, bottom-left, and bottom-right, respectively). Let's take the fractional parts of *tx* and *ty* and call them *fx* and *fy*. Figure 14-9 shows *C*, the exact position described by (*tx*, *ty*), surrounded by the texels at integer coordinates, and its distance to them.

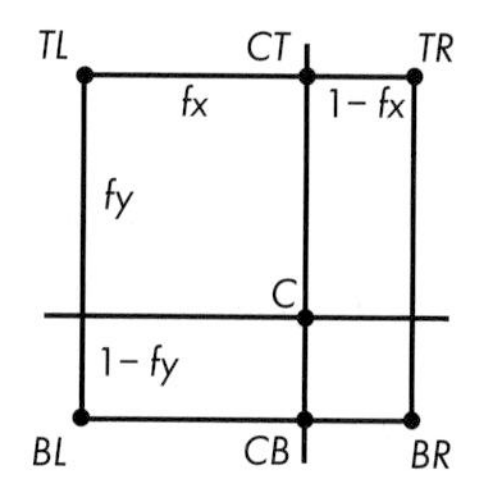

Figure 14-9: We linearly interpolate a color at C from the four texels that surround it.

First, we linearly interpolate the color at *CT*, which is between *TL* and *TR*:

$$CT = (1 - fx) \cdot TL + fx \cdot TR$$

Note that the weight for *TR* is *fx*, not $(1 - fx)$. This is because as *fx* becomes closer to 1.0, we want *CT* to become closer to *TR*. Indeed, if $fx = 0.0$, then $CT = TL$, and if $fx = 1.0$, then $CT = TR$.

We can compute *CB*, between *TL* and *TR*, in a similar way:

$$CB = (1 - fx) \cdot BL + fx \cdot BR$$

Finally, we compute C, linearly interpolating between CT and CB:

$$C = (1 - fy) \cdot BT + fy \cdot CB$$

In pseudocode, we can write a function to get the interpolated color corresponding to a fractional texel:

```
GetTexel(texture, tx, ty) {
  fx = frac(tx)
  fy = frac(ty)
  tx = floor(tx)
  ty = floor(ty)

  TL = texture[tx][ty]
  TR = texture[tx+1][ty]
  BL = texture[tx][ty+1]
  BR = texture[tx+1][ty+1]

  CT = fx * TR + (1 - fx) * TL
  CB = fx * BR + (1 - fx) * BL

  return fy * CB + (1 - fy) * CT
}
```

This function uses `floor()`, which rounds a number down to the nearest integer, and `frac()`, which returns the fractional part of a number, and can be defined as `x - floor(x)`.

This technique is called *bilinear filtering* (because we're doing linear interpolation twice, once in each dimension).

Mipmapping

Let's consider the opposite situation, rendering an object from far away. In this case, the texture has many more texels than the triangle has pixels. It might be less evident why this is a problem, so we'll use a carefully chosen situation to illustrate.

Consider a square texture in which half the pixels are black and half the pixels are white, laid out in a checkerboard pattern (Figure 14-10).

Figure 14-10: A black-and-white checkerboard texture

Suppose we map this texture onto a square in the viewport such that when it's drawn on the canvas, the width of the square in pixels is exactly half the width of the texture in texels. This means that only one-quarter of the texels will actually be used.

We'd intuitively expect the square to look gray. However, given the way we're doing texture mapping, we might be unlucky and get all the white pixels, or all the black pixels. It's true that we might be lucky and get a 50/50 combination of black and white pixels, but the 50-percent gray we expect is not guaranteed. Take a look at Figure 14-11, which shows the unlucky case.

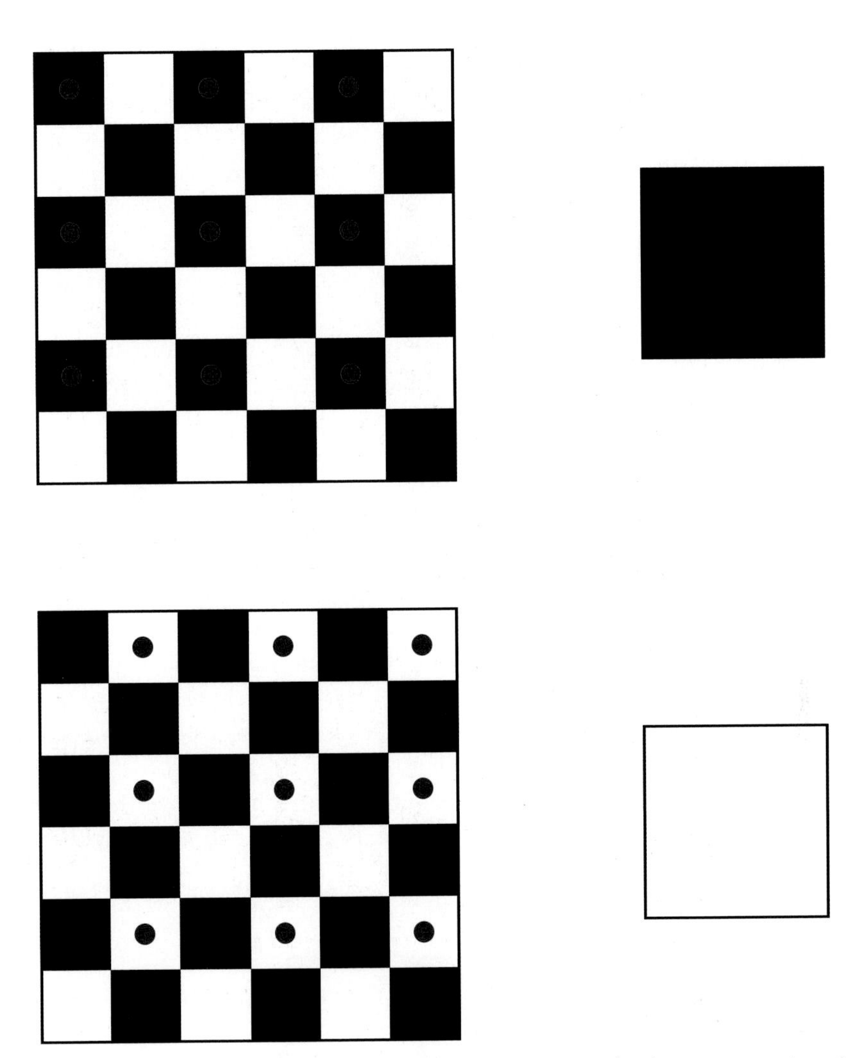

Figure 14-11: Mapping a big texture on a small object can lead to unexpected results, depending on which texels happen to be selected.

How to fix this? Each pixel of the square represents, in some sense, a 2×2 texel area of the texture, so we could compute the average color of that area and use that color for the pixel. Averaging black and white pixels would give us the gray we are looking for.

However, this can get very computationally expensive very fast. Suppose the square is even farther away, so that it's one-tenth of the texture width. This means every pixel in the square represents a 10×10 texel area of the texture. We'd have to compute the average of 100 texels for every pixel we want to render!

Fortunately, this is one of those situations where we can replace a lot of computation with a bit of extra memory. Let's go back to the initial situation, where the square was half the width of the texture. Instead of computing the average of the four texels we want to render for every pixel again and again, we could *precompute* a texture of half the original size, where every texel in the half-size texture is the average of the corresponding four texels in the original texture. Later, when the time comes to render a pixel, we can just look up the texel in this smaller texture, or even apply bilinear filtering as described in the previous section.

This way, we get the better rendering quality of averaging four pixels, but at the computational cost of a single texture lookup. This does require a bit of preprocessing time (when loading a texture, for example) and a bit more memory (to store the full-size and half-size textures), but in general it's a worthwhile trade-off.

What about the 10× size scenario we discussed above? We can take this technique further and also precompute one-quarter-, one-eighth-, and one-sixteenth-size versions of the original texture (down to a 1×1 texture if we wanted to). Then, when rendering a triangle, we'd use the texture whose scale best matches its size and get all the benefits of averaging hundreds, if not thousands, of pixels at no extra runtime cost.

This powerful technique is called *mipmapping*. The name is derived from the Latin expression *multum in parvo*, which means "much in little."

Computing all these smaller-scale textures does come at a memory cost, but it's surprisingly smaller than you might think.

Say the original area of the texture, in texels, is A, and its width is w. The width of the half-width texture is $\frac{w}{2}$, but it requires only $\frac{A}{4}$ texels; the quarter-width texture requires $\frac{A}{16}$ texels; and so on. Figure 14-12 shows the original texture and the first three reduced versions.

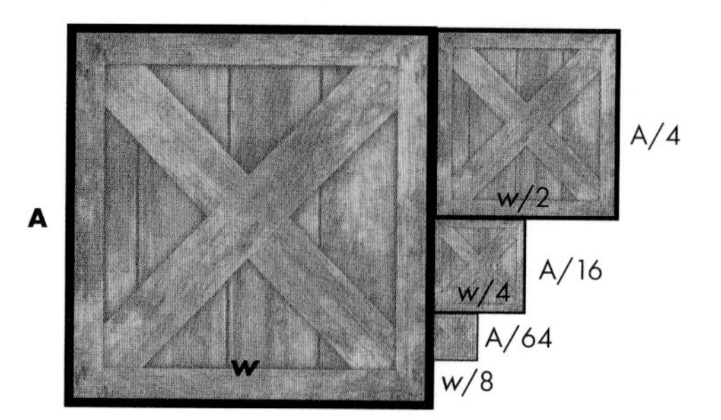

Figure 14-12: A texture and its progressively smaller mipmaps

We can express the sum of the texture sizes as an infinite series:

$$A + \frac{A}{4} + \frac{A}{16} + \frac{A}{64} + \ldots = \sum_{n=0}^{\infty} \frac{A}{4^n}$$

This series converges to $A \cdot 4/3$, or $A \cdot 1.3333$, meaning that all the smaller textures down to 1×1 texel only take one-third more space than the original texture.

Trilinear Filtering

Let's take this one step further. Imagine an object far away from the camera. We render it using the mipmap level most appropriate for its size.

Now imagine the camera moves toward the object. At some point, the choice of the most appropriate mipmap level will change from one frame to the next, and this will cause a subtle but noticeable difference.

When choosing a mipmap level, we choose the one that most closely matches the relative size of the texture and the square. For example, for the square that was 10 times smaller than the texture, we might choose the mipmap level that is 8 times smaller than the original texture, and apply bilinear filtering on it. However, we could also consider the *two* mipmap levels that most closely match the relative size (in this case, the ones 8 and 16 times smaller) and linearly interpolate between them, depending on the "distance" between the mipmap size ratio and the actual size ratio.

Because the colors that come from each mipmap level are bilinearly interpolated and we apply another linear interpolation on top, this technique is called *trilinear filtering*.

Summary

In this chapter, we have given our rasterizer a massive jump in quality. Before this chapter, each triangle could have a single color; now we can draw arbitrarily complex images on them.

We have also discussed how to make sure the textured triangles look good, regardless of the relative size of the triangle and the texture. We presented bilinear filtering, mipmapping, and trilinear filtering as solutions to the most common causes of low-quality textures.

15

EXTENDING THE RASTERIZER

We'll conclude this second part of the book the same way we concluded the first one: with a set of possible extensions to the rasterizer we've developed in the preceding chapters.

Normal Mapping

In Chapter 13, we saw how the normal vectors of a surface have a big impact on its appearance. For example, the right choice of normals can make a faceted object look smoothly curved; this is because the right choice of normals changes the way light interacts with the surface, which in turn changes the way our brain guesses the shape of the object. Unfortunately, there's not much more we can do by interpolating normals beyond making surfaces look smoothly curved.

In Chapter 14, we saw how we could add fake detail to a surface by "painting" on it. This technique, called texture mapping, gives us much finer-grained control over the appearance of a surface. However, texture mapping doesn't change the shape of the triangles—they're still flat.

Normal mapping combines both ideas. We can use normals to change the way light interacts with a surface and thus change the apparent shape of the surface; we can use attribute mapping to assign different values of an attribute to different parts of a triangle. By combining the two ideas, normal mapping lets us define surface normals at the pixel level.

To do this, we associate a *normal map* to each triangle. A normal map is similar to a texture map, but its elements are normal vectors instead of colors. At rendering time, instead of computing an interpolated normal like Phong shading does, we use the normal map to get a normal vector for the specific pixel we're rendering, in the same way that texture mapping gets a color for that specific pixel. Then we use this vector to compute lighting at that pixel.

Figure 15-1 shows a flat surface with a texture map applied, and the effects of different light directions when a normal map is also applied.

(a) No normal map

(b) Normal map plus light from the left

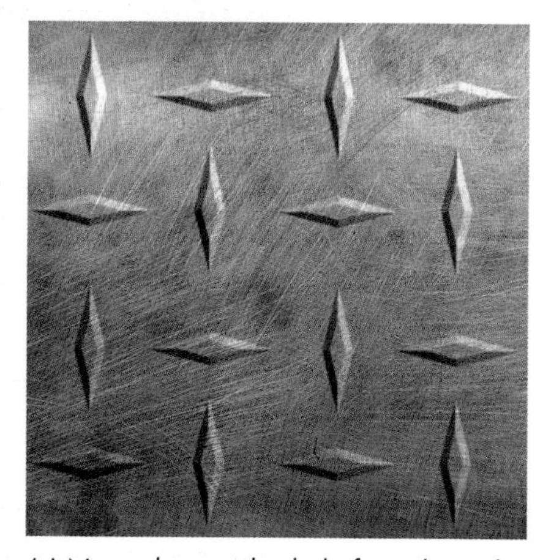

(c) Normal map plus light from the right

Figure 15-1: The effect of normal mapping over flat geometry

All three images in Figure 15-1 are renders of a flat square (that is, two triangles) with a texture, as seen in (a). When we add a normal map and the appropriate per-pixel shading, we create the illusion of extra geometrical detail. In (b) and (c), the shading of the diamonds depends on the direction of the incident light, and our brain interprets this as the diamonds having volume.

There are a couple of practical considerations to keep in mind. First, the orientations of the vectors in the normal map are relative to the surface of the triangle they apply to. The coordinate system used for this is called *tangent space*, where two of the axes (usually *X* and *Z*) are tangent to (that is, embedded in) the surface and the remaining vector is perpendicular to the surface. At rendering time, the normal vector of the triangle, expressed in camera space, is modified according to the vector in the normal map to obtain a final normal vector that can be used for the illumination equations. This makes a normal map independent of the position and orientation of the object in the scene.

Second, a very popular way to encode normal maps is as textures, mapping the values of *X*, *Y*, and *Z* to *R*, *G*, and *B* values. This gives normal maps a very characteristic purple-ish appearance, because purple, a combination of red and blue but no green, encodes flat areas of the surface. Figure 15-2 shows the normal map used in the examples in Figure 15-1.

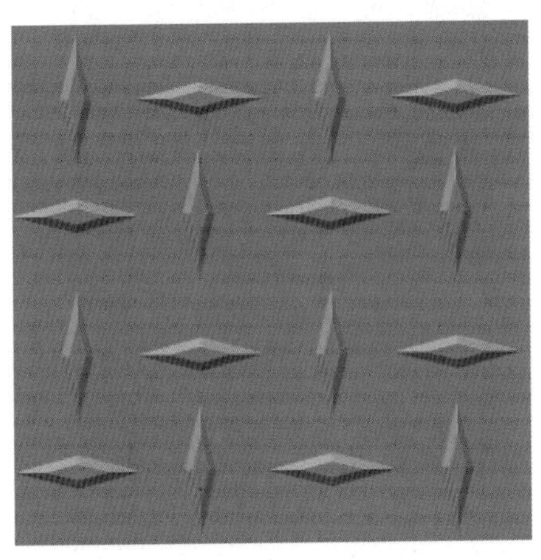

Figure 15-2: The normal map used for the examples in Figure 15-1, encoded as a RGB texture

While this technique can drastically improve the perceived complexity of surfaces in a scene, it's not without limitations. For example, since flat surfaces remain flat, it can't change the silhouette of an object. For the same reason, the illusion breaks down when a normal-mapped surface is viewed from an extreme angle or up close, or when the features represented by the normal map are too big compared to the size of the surface. This technique is better suited to subtle detail, such as pores on the skin, the pattern on a stucco wall, or the irregular appearance of an orange peel. For this reason, the technique is also known as *bump mapping*.

Environment Mapping

One of the most striking characteristics of the raytracer we developed is the ability to show objects reflecting one another. It is possible to create a relatively convincing, but somewhat fake, implementation of reflections in our rasterizer.

Imagine we have a scene representing a room in a house, and we want to render a reflective object placed in the middle of the room. For each pixel representing the surface of that object, we know the 3D coordinates of the point it represents, the surface normal at that point, and, since we know the position of the camera, we can also compute the view vector to that point. We could reflect the view vector with respect to the surface normal to obtain a reflection vector, just like we did in Chapter 4.

At this point, we want to know the color of the light coming from the direction of the reflection vector. If this were a raytracer, we'd just trace a ray in that direction and find out. However, this is not a raytracer. What to do?

Environment mapping provides one possible answer to this question. Suppose that before rendering the objects inside the room, we place a camera in the middle of it and render the scene six times—once for each perpendicular direction (up, down, left, right, front, back). You can imagine the camera is inside an imaginary cube, and each side of the cube is the viewport of one of these renders. We keep these six renders as textures. We call this set of six textures a *cube map*, which is why this technique is also called *cube mapping*.

Then we render the reflective object. When we get to the point of needing a reflected color, we can use the direction of the reflected vector to choose one of the textures of the cube map, and then a texel of that texture to get an approximation of the color seen in that direction—all without tracing a single ray!

This technique has some drawbacks. The cube map captures the appearance of a scene from a single point. If the reflective object we're rendering isn't located at that point, the position of the reflected objects won't fully match what we would expect, so it will become clear that this is just an approximation. This would be especially noticeable if the reflective object were to move within the room, because the reflected scene wouldn't change as the object moves.

This limitation also suggests the best applications for the technique: if the "room" is big enough and distant enough from the object—that is, if the movement of the object is small with respect to the size of the room—the difference between the true reflection and the pre-rendered environment maps can go unnoticed. For example, this would work very well for a scene representing a reflective spaceship in deep space, since the "room" (the distant stars and galaxies) is infinitely far away for all practical purposes.

Another drawback is that we're forced to split the objects in the scene into two categories: static objects that are part of the "room," which are seen in reflections, and dynamic objects that can be reflective. In some cases, this might be clear (walls and furniture are part of the room; people aren't), but even then, dynamic objects wouldn't be reflected on other dynamic objects.

A final drawback worth mentioning is related to the resolution of the cube maps. Whereas in the raytracer we could trace very precise reflections, in this case we need to make a trade-off between accuracy (higher-resolution cube map textures produce sharper reflections) and memory consumption (higher-resolution cube map textures require more memory). In practice, this means that environment maps won't produce reflections that are as sharp as true raytraced reflections, especially when looking at reflective objects up close.

Shadows

The raytracer we developed featured geometrically correct, very well-defined shadows. These were a very natural extension to the core algorithm. The architecture of a rasterizer makes it slightly more complex to implement shadows, but not impossible.

Let's start by formalizing the problem we're trying to solve. In order to render shadows correctly, every time we compute the illumination equation for a pixel and a light, we need to know whether the pixel is actually illuminated by the light or whether it's in the shadow of an object with respect to that light.

With the raytracer, we could answer this question by tracing a ray from the surface to the light; in the rasterizer, we don't have such a tool, so we'll have to take a different approach. Let's explore two different approaches.

Stencil Shadows

Stencil shadows is a technique to render shadows with very well-defined edges (imagine the shadows cast by objects on a very sunny day). These are often called *hard shadows*.

Our rasterizer renders the scene in a single *pass*; it goes over every triangle in the scene and renders it on the canvas, computing the full illumination equation every time (on a per-triangle, per-vertex, or per-pixel basis, depending on the shading algorithm). At the end of this process, the canvas contains the final render of the scene.

We'll start by modifying the rasterizer to render the scene in several *passes*, one for each light in the scene (including the ambient light). Like before, each pass goes over every triangle, but it computes the illumination equation taking into account only the light associated with that pass.

This gives us a set of images of the scene illuminated by each light separately. We can *compose* them together—that is, add them pixel by pixel—giving us the final render of the scene. This final image is identical to the image produced by the single-pass version. Figure 15-3 shows three light passes and the final composite for our reference scene.

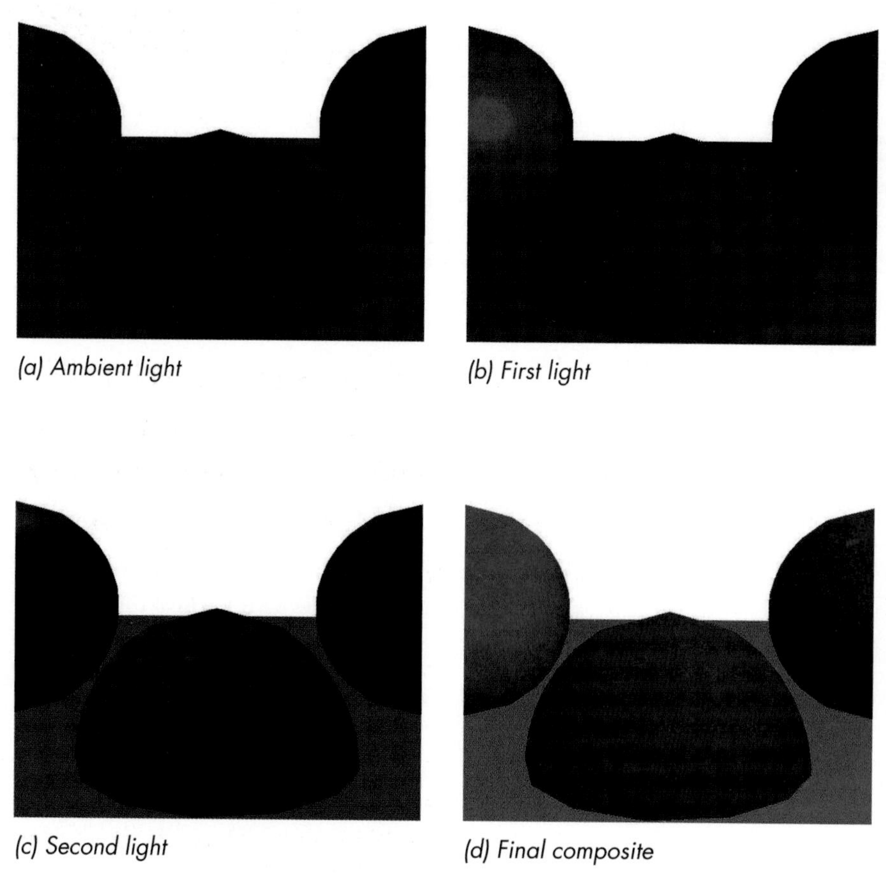

(a) Ambient light *(b) First light*

(c) Second light *(d) Final composite*

Figure 15-3: A scene rendered using one pass per light

This lets us simplify our goal of "rendering a scene with shadows from multiple lights" into "rendering a scene with shadows from a single light, many times." Now we need to find a way to render a scene illuminated by a single light, while leaving the pixels that are in shadow from that light completely black.

For this, we introduce the *stencil buffer*. Like the depth buffer, it has the same dimensions as the canvas, but its elements are integers. We can use it as a stencil for rendering operations, for example, modifying our rendering code to draw a pixel on the canvas only if the corresponding element in the stencil buffer has a value of zero.

If we can set up the stencil buffer in such a way that illuminated pixels have a value of zero and pixels in shadow have a nonzero value, we can use it to draw only the pixels that are illuminated.

Creating Shadow Volumes

To set up the stencil buffer, we use something called *shadow volumes*. A shadow volume is a 3D polygon "wrapped" around the volume of space that's in shadow from a light.

We construct a shadow volume for each object that might cast a shadow on the scene. First, we determine which edges are part of the silhouette of the object; these are the edges between front-facing and back-facing triangles (we can use the dot product to classify the triangles, like we did for the back-face culling technique in Chapter 12). Then, for each of these edges, we extrude them away from the direction of the light, all the way to infinity—or, in practice, to a really big distance beyond the scene.

This gives us the "sides" of the shadow volume. The "front" of the volume is made by the front-facing triangles of the object itself, and the "back" of the volume can be computed by creating a polygon whose edges are the "far" edges of the extruded sides.

Figure 15-4 shows the shadow volume created this way for a cube with respect to a point light.

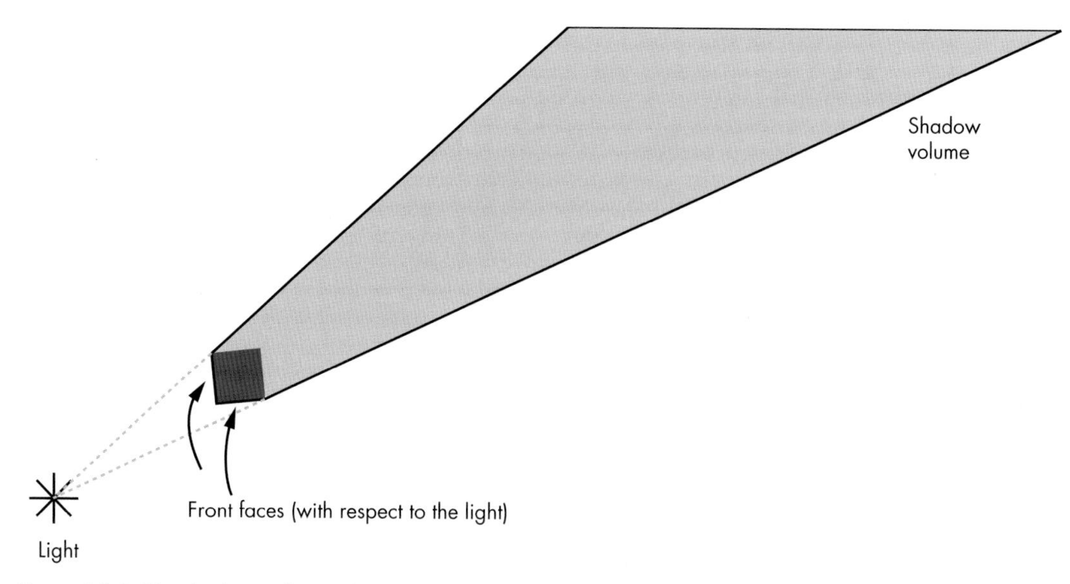

Figure 15-4: The shadow volume of a cube with respect to a point light

Next, we'll see how to use the shadow volumes to determine which pixels in the canvas are in shadow with respect to a light.

Counting Shadow Volume–Ray Intersections

Imagine a ray starting from the camera and going into the scene until it hits a surface. Along the way, it might enter and leave any number of shadow volumes.

We can keep track of this with a counter that starts at zero. Every time the ray enters a shadow volume, we increment the counter; every time it leaves, we decrement it. We stop when the ray hits a surface and look at the counter. If it's zero, it means the ray entered as many shadow volumes as it left, so the point must be illuminated; if it's not zero, it means the ray is inside at least one shadow volume, so the point must be in shadow. Figure 15-5 shows a few examples of this.

However, this only works if the camera itself is not inside a shadow volume! If a ray starts inside the shadow volume and doesn't leave it before hitting the surface, our algorithm would incorrectly conclude that it's illuminated.

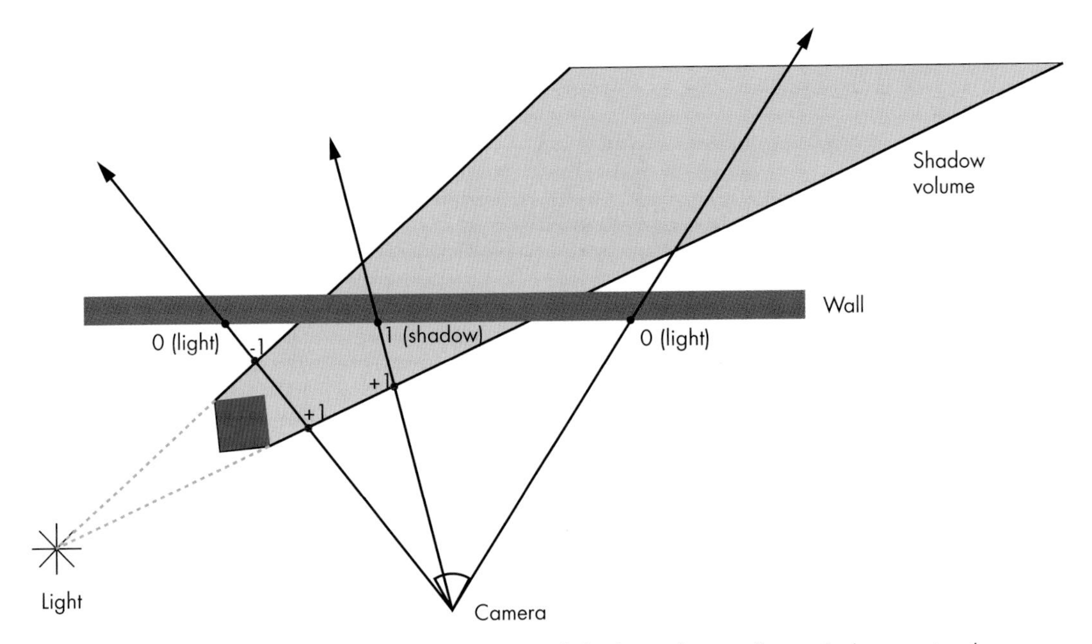

Figure 15-5: Counting the intersections between rays and shadow volumes tells us whether a point along the ray is illuminated or in shadow.

We could check for this condition and adjust the counter accordingly, but counting how many shadow volumes a point is inside of is an expensive operation. Fortunately, there's a way to overcome this limitation that is simpler and cheaper, if somewhat counter-intuitive.

Rays are infinite, but shadow volumes aren't. This means a ray always starts and ends outside a shadow volume. This, in turn, means that a ray always enters a shadow volume as many times as it leaves it; the counter for the entire ray must always be zero.

Suppose we keep track of the intersections between the ray and the shadow volume *after* the ray hits the surface. If the counter has a value of zero, then the value must also be zero *before* the ray hits the surface. If the counter has a nonzero value, it must have the opposite value on the other side of the surface.

This means counting intersections between the ray and the shadow volume before the ray hits the surface is equivalent to counting the intersections after it—but in this case, we don't have to worry about the position of the camera! Figure 15-6 shows how this technique always produces correct results.

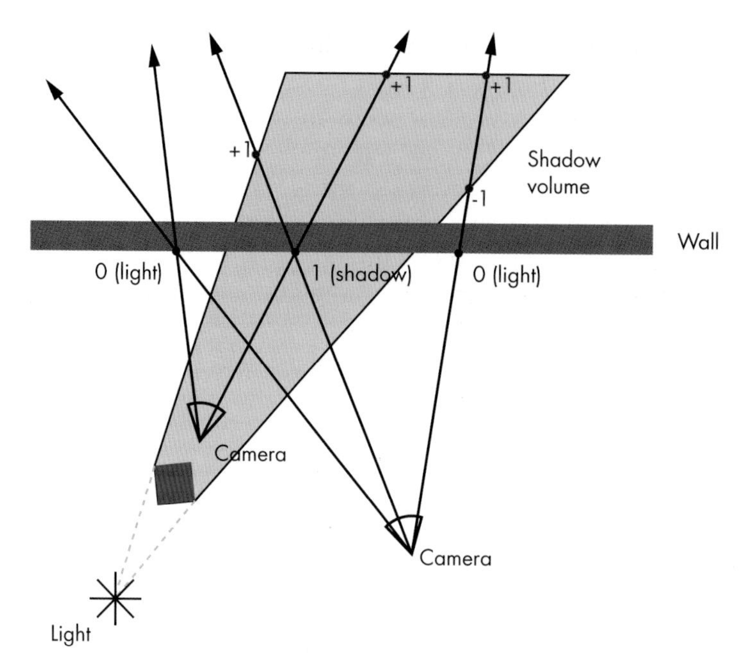

Figure 15-6: The counters have a value of zero for points that receive light, and a nonzero value for points that are in shadow, regardless of whether the camera is inside or outside the shadow volume.

Setting up the Stencil Buffer

We're working with a rasterizer, not with a raytracer, so we need to find a way to keep these counters without actually computing any intersections between rays and shadow volumes. We can do this by using the stencil buffer.

First, we render the scene as illuminated only by the ambient light. The ambient light casts no shadows, so we can do this without any changes to the rasterizer. This gives us one of the images we need to compose the final render, but it also gives us depth information for the scene, as seen from the camera, contained in the depth buffer. We need to keep this depth buffer for the next steps.

Next, for each light, we follow these steps:

1. "Render" the back faces of the shadow volumes to the stencil buffer, incrementing its value whenever the pixel *fails* the depth buffer test. This counts the number of times the ray leaves a shadow volume after hitting the closest surface.
2. "Render" the front faces of the shadow volumes to the stencil buffer, decrementing its value whenever the pixel *fails* the depth buffer test. This counts the number of times the ray enters a shadow volume after hitting the closest surface.

Note that during the "rendering" steps, we're only interested in modifying the stencil buffer; there's no need to write pixels to the canvas, and therefore no need to calculate illumination or texturing. We also don't write to the depth buffer, because the sides of the shadow volumes aren't actually physical objects in the scene. Instead, we use the depth buffer we computed during the ambient lighting pass.

After doing this, the stencil buffer has zeros for the pixels that are illuminated and other values for the pixels that are in shadow. So we render the scene normally, illuminated by the single light corresponding to this pass, calling `PutPixel` only on the pixels where the stencil buffer has a value of zero.

Repeating this process for every light, we end up with a set of images corresponding to the scene illuminated by each of the lights, with shadows correctly taken into account. The final step is to compose all the images into a final render of the scene by adding them together pixel by pixel.

The idea of using the stencil buffer to render shadows dates back to the early 1990s, but the first implementations had several drawbacks. The depth-fail variant described here was independently discovered several times during 1999 and 2000, most notably by John Carmack while working on *Doom 3*, which is why this variant is also known as *Carmack's Reverse*.

Shadow Mapping

The other well-known technique to render shadows in a rasterizer is called *shadow mapping*. This renders shadows with less defined edges (imagine the shadows cast by objects on a cloudy day). These are often called *soft shadows*.

To reiterate, the question we're trying to answer is, given a point on a surface and a light, does the point receive illumination from that light? This is equivalent to determining whether there's an object between the light and the point.

With the raytracer, we traced a ray from the point to the light. In some sense, we're asking whether the point can "see" the light, or, equivalently, whether the light can "see" the point.

This leads us to the core idea of shadow mapping. We render the scene from the point of view of the light, preserving the depth buffer. Similar to how we created the environment maps we described above, we render the scene six times and end up with six depth buffers. These depth buffers, which we call *shadow maps*, let us determine the distance to the closest surface the light can "see" in any given direction.

The situation is slightly more complicated for directional lights, because these have no position to render from. Instead, we need to render the scene from a *direction*. This requires using an *orthographic projection* instead of our usual perspective projection. With perspective projection and point lights, every ray starts from a point; with orthographic projection and directional lights, every ray is parallel to each other, sharing the same direction.

When we want to determine whether a point is in shadow or not, we compute the distance and the direction from the light to the point. We use the direction to look up the corresponding entry in the shadow map. If this depth value is smaller than the distance from the point to the light, it means there's a surface that is closer to the light than the point we're illuminating, and therefore the point is in the shadow of that surface; otherwise, the light can "see" the point unobstructed, so the point is illuminated by the light.

Note that the shadow maps have a limited resolution, usually lower than the canvas. Depending on the distance and the relative orientation of the point and the light, this might cause the shadows to look blocky. To avoid this, we can sample the depth of the surrounding depth entries as well and determine whether the point lies on the edge of a shadow (as evidenced by a depth discontinuity in the surrounding entries). If this is the case, we can use a technique similar to bilinear filtering, as we did in Chapter 14, to come up with a value between 0.0 and 1.0 representing *how much* the point is visible from the light and multiply it by the illumination of the light; this gives the shadows created by shadow mapping their characteristic blurry appearance. Other ways to avoid the blocky appearance involve sampling the shadow map in different ways—for example, look into *percentage closer filtering*.

Summary

Like in Chapter 5, this chapter briefly introduced several ideas you can explore by yourself. These extend the rasterizer developed over the previous chapters to bring its features closer to those of the raytracer, while retaining its speed advantage. There's always a trade-off, and in this case it comes in the form of less accurate results or increased memory consumption, depending on the algorithm.

AFTERWORD

Congratulations! You now have a good understanding of how 3D rendering works. You've created a raytracer and a rasterizer and gained a good conceptual understanding of the algorithms and math that power them.

However, as I explained in the introduction, it's impossible to cover the entirety of 3D rendering in a single book. Here's a few topics you might want to explore on your own to expand your horizons:

Global illumination, including radiosity and path tracing Find out how deep the "ambient light" rabbit hole goes!

Physically based rendering Illumination and shading models that don't just look good, but model real-life physics.

Voxel rendering Think Minecraft, or MRI scans in hospitals.

Level-of-detail algorithms This includes offline and dynamic mesh simplification, impostors, and billboards. These algorithms are how we efficiently render forests with billions of plants, crowds of millions of people, or extremely detailed 3D models.

Acceleration structures This includes binary space partition trees, k-d trees, quadtrees, and octrees. These structures help efficiently render massive scenes, such as an entire city.

Terrain rendering How to efficiently render a terrain model that might be as big as a country yet have human-scale detail.

Atmospheric effects and particle systems Fog, rain, and smoke, but also some less intuitive materials like grass and hair.

Image-based lighting Similar to environment mapping, but for diffuse lighting.

High dynamic range, gamma correction The color representation rabbit hole also goes deep.

Caustics Also known as "the moving white patterns at the bottom of the swimming pool."

Procedural generation of textures and models Add more variety and possibly infinitely big scenes.

Hardware acceleration Using OpenGL, Vulkan, DirectX, and others to run graphics algorithms on GPUs.

Of course, there are many other topics, and that's just 3D rendering! Computer graphics is an even broader subject. Here are some areas you might want to investigate:

Font rendering This is surprisingly more complex than you might think.

Image compression How to store images in the least amount of space.

Image processing (such as transforming and filtering) Think Instagram filters.

Image recognition Is that a dog or a cat?

Curve rendering, including Bezier curves and splines Find out what these weird arrows on the curves of your favorite drawing program really are!

Computational photography How does the camera on your phone take such good pictures with almost no light?

Image segmentation Before you can "blur the background" of your video call, you need to determine which pixels are background and which aren't.

Congratulations again on taking your first step into the world of computer graphics. Now you get to choose where to go next!

LINEAR ALGEBRA

This appendix serves as a cheat sheet for linear algebra. The subject is presented as a set of tools, their properties, and what you can use them for. If you're interested in the theory behind all this, you can pick up any introductory linear algebra textbook.

The focus here is exclusively on 2D and 3D algebra, as that's what's required in this book.

Points

A *point* represents a position within a coordinate system.

We represent a point as a sequence of numbers between parentheses—for example, $(4, 3)$. We refer to points using capital letters, such as P or Q.

Each of the numbers in the point's sequence is called a *coordinate*. The number of coordinates is the point's *dimension*. A point with two coordinates is called two-dimensional, or 2D.

The order of the numbers is important; $(4, 3)$ is not the same as $(3, 4)$. By convention, the coordinates are called x and y in 2D, and x, y, and z in

3D; so the point (4, 3) has an x coordinate of 4 and a y coordinate of 3. Figure A-1 shows P, a 2D point with coordinates (4, 3).

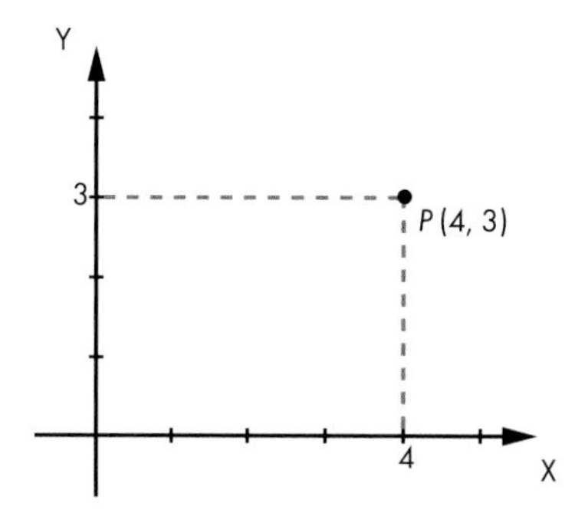

Figure A-1: The 2D point P *has coordinates (4, 3).*

We can also refer to specific coordinates of a point using a subscript, like P_x or Q_y. So the point P can also be written as (P_x, P_y, P_z) when convenient.

Vectors

A *vector* represents the difference between two points. Intuitively, imagine a vector as an arrow that connects a point to another point; alternatively, think of it as the instructions to get from one point to another.

Representing Vectors

We represent a vector as a set of numbers between parentheses, and refer to them using a capital letter. This is the same representation we use for points, so we add a small arrow on top to remember they're vectors and not points. For example, (2, 1) is a vector, which we might decide to call $\vec{A}$. Figure A-2 shows two equal vectors, $\vec{A}$ and $\vec{B}$.

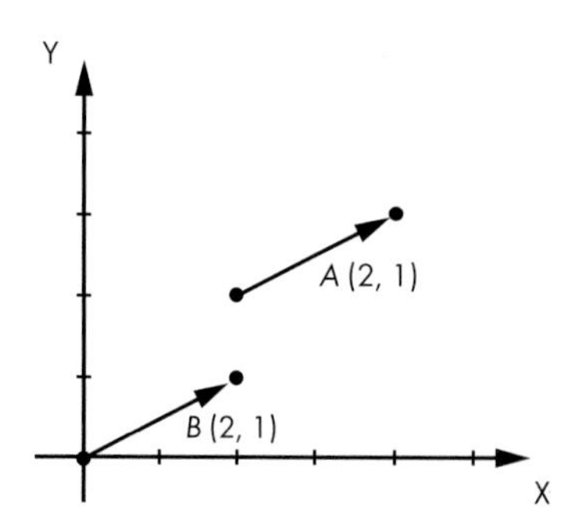

Figure A-2: The vectors $\vec{A}$ and $\vec{B}$ are equal. Vectors don't have a position.

Despite sharing their representation with points, vectors don't represent or have a position; they are, after all, the *difference* between two positions.

When you have a diagram like Figure A-2, you have to draw vectors somewhere; but the vectors $\vec{A}$ and $\vec{B}$ are equal, because they represent the same displacement.

In addition, the point (2, 1) and the vector (2, 1) are unrelated. Sure, the vector (2, 1) goes from (0, 0) to (2, 1), but it's equally true that it goes from, say, (5, 5) to (7, 6).

Vectors are characterized by their *direction* (the angle in which they point) and their *magnitude* (how long they are).

The direction can be further decomposed into *orientation* (the slope of the line they're on) and *sense* (which of the possible two ways along that line they point). For example, a vector pointing right and a vector pointing left both have the same horizontal orientation, but they have the opposite sense. However, we don't make this distinction anywhere in this book.

Vector Magnitude

You can compute the magnitude of a vector from its coordinates. The magnitude is also called the *length* or *norm* of the vector. It's denoted by putting the vector between vertical pipes, as in $|\vec{V}|$, and it's computed as follows:

$$|\vec{V}| = \sqrt{V_x{}^2 + V_y{}^2 + V_z{}^2}$$

A vector with a magnitude equal to 1.0 is called a *unit vector*.

Point and Vector Operations

Now that we've defined points and vectors, let's explore what we can do with them.

Subtracting Points

A vector is the difference between two points. In other words, you can subtract two points and get a vector:

$$\vec{V} = P - Q$$

In this case, you can think of $\vec{V}$ as "going" from Q to P, as in Figure A-3.

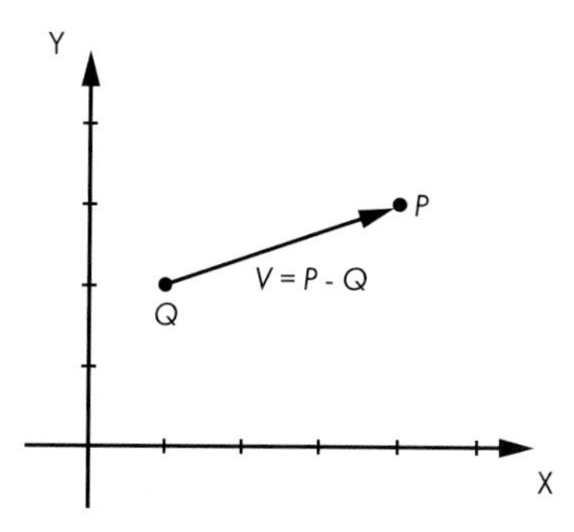

Figure A-3: The vector $\vec{V}$ *is the difference between* P *and* Q*.*

Algebraically, you subtract each of the coordinates separately:

$$(V_x, V_y, V_z) = (P_x, P_y, P_z) - (Q_x, Q_y, Q_z) = (P_x - Q_x, P_y - Q_y, P_z - Q_z)$$

Adding a Point and a Vector

We can rewrite the equation above coordinate by coordinate:

$$V_x = P_x - Q_x$$

$$V_y = P_y - Q_y$$

$$V_z = P_z - Q_z$$

These are just numbers, so all the usual rules apply. This means you can do this:

$$Q_x + V_x = P_x$$

$$Q_y + V_y = P_y$$

$$Q_z + V_z = P_z$$

And grouping the coordinates again,

$$Q + \vec{V} = P$$

In other words, you can add a vector to a point and get a new point. This makes intuitive and geometric sense; given a starting position (a point) and a displacement (a vector), you end up in a new position (another point). Figure A-4 presents an example.

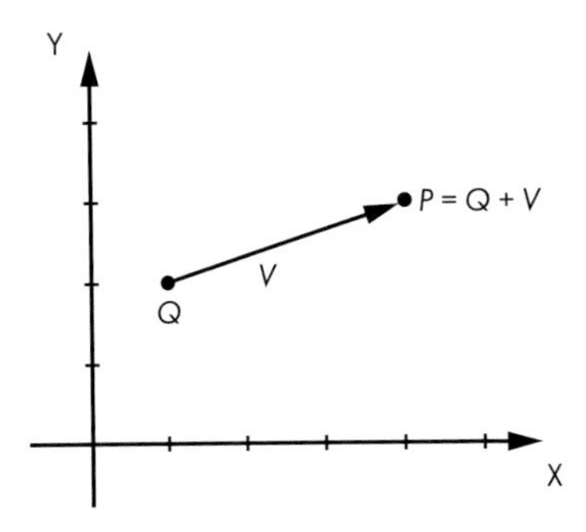

Figure A-4: Adding $\vec{V}$ *to* Q *gives us* P.

Adding Vectors

You can add two vectors. Geometrically, imagine putting one vector "after" another, as in Figure A-5.

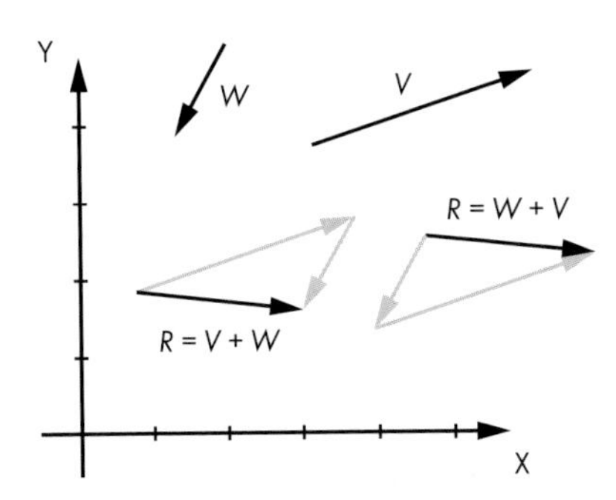

Figure A-5: Adding two vectors. Addition is commutative. Remember, vectors don't have a position.

As you can see, vector addition is commutative—that is, the order of the operands doesn't matter. In the diagram, we can see that $\vec{V} + \vec{W} = \vec{W} + \vec{V}$.

Algebraically, you add the coordinates individually:

$$\vec{V} + \vec{W} = (V_x, V_y, V_z) + (W_x, W_y, W_z) = (V_x + W_x, V_y + W_y, V_z + W_z)$$

Multiplying a Vector by a Number

You can multiply a vector by a number. This is called the *scalar product*. This makes the vector shorter or longer, as you can see in Figure A-6.

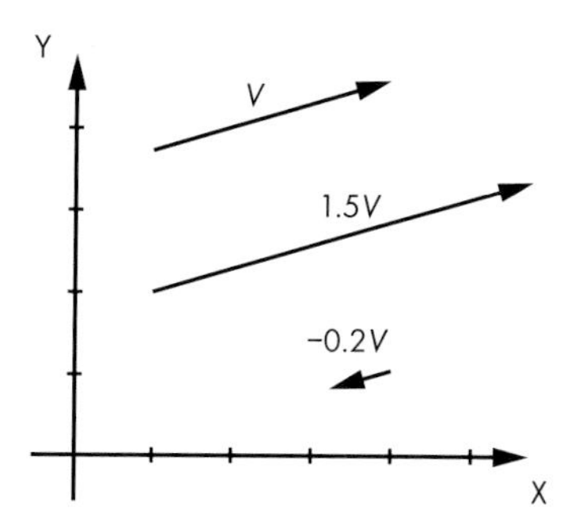

Figure A-6: Multiplying a vector by a number

If the number is negative, the vector will point the other way; this means it changes its sense and therefore its direction. But multiplying a vector by a number never changes its orientation—that is, it will remain along the same line.

Algebraically, you multiply the coordinates individually:

$$k \cdot \vec{V} = k \cdot (V_x, V_y, V_z) = (k \cdot V_x, k \cdot V_y, k \cdot V_z)$$

You can also divide a vector by a number. Just like with numbers, dividing by k is equivalent to multiplying by $\frac{1}{k}$. As usual, division by zero doesn't work.

One of the applications of vector multiplication and division is to *normalize* a vector—that is, to turn it into a unit vector. This changes the magnitude of the vector to 1.0, but doesn't change its other properties. To do this, we just need to divide the vector by its length:

$$\vec{V}_{normalized} = \frac{\vec{V}}{|\vec{V}|}$$

Multiplying Vectors

You can multiply a vector by another vector. Interestingly, there are many ways in which you can define an operation like this. We're going to focus on two kinds of multiplication that are useful to us: the dot product and the cross product.

Dot Product

The *dot product* between two vectors (also called the *inner product*) gives you a number. It's expressed using the dot operator, as in $\vec{V} \cdot \vec{W}$. It's also written between angle braces, as in $\langle \vec{V}, \vec{W} \rangle$.

Algebraically, you multiply the coordinates individually and add them:

$$\langle \vec{V}, \vec{W} \rangle = \langle (V_x, V_y, V_z), (W_x, W_y, W_z) \rangle = V_x \cdot W_x + V_y \cdot W_y + V_z \cdot W_z$$

Geometrically, the dot product of $\vec{V}$ and $\vec{W}$ is related to their lengths and to the angle α between them. The exact formula neatly ties together linear algebra and trigonometry:

$$\langle \vec{V}, \vec{W} \rangle = |\vec{V}| \cdot |\vec{W}| \cdot \cos(\alpha)$$

Either of these formulas help us see that the dot product is commutative (that is, $\langle \vec{V}, \vec{W} \rangle = \langle \vec{W}, \vec{V} \rangle$) and that it's distributive with respect to a scalar product (that is, $k \cdot \langle \vec{V}, \vec{W} \rangle = \langle k \cdot \vec{V}, \vec{W} \rangle$).

An interesting consequence of the second formula is that if $\vec{V}$ and $\vec{W}$ are perpendicular, then $\cos(\alpha) = 0$ and therefore $\langle \vec{V}, \vec{W} \rangle$ is also zero. If $\vec{V}$ and $\vec{W}$ are unit vectors, then $\langle \vec{V}, \vec{W} \rangle$ is always between −1.0 and 1.0, with 1.0 meaning they're equal and −1.0 meaning they're opposite.

The second formula also suggests the dot product can be used to calculate the angle between two vectors:

$$\alpha = \cos^{-1}\left(\frac{\langle \vec{V}, \vec{W} \rangle}{|\vec{V}| \cdot |\vec{W}|} \right)$$

Note that the dot product of a vector with itself, $\langle \vec{V}, \vec{V} \rangle$, reduces to the square of its length:

$$\langle \vec{V}, \vec{V} \rangle = V_x^2 + V_y^2 + V_z^2 = |\vec{V}|^2$$

This suggests another way to compute the length of a vector, as the square root of its dot product with itself:

$$|\vec{V}| = \sqrt{\langle \vec{V}, \vec{V} \rangle}$$

Cross Product

The cross product between two vectors gives you another vector. It's expressed using the cross operator, as in $\vec{V} \times \vec{W}$.

The cross product of two vectors is a vector perpendicular to both of them. In this book we only use the cross product on 3D vectors, shown in Figure A-7.

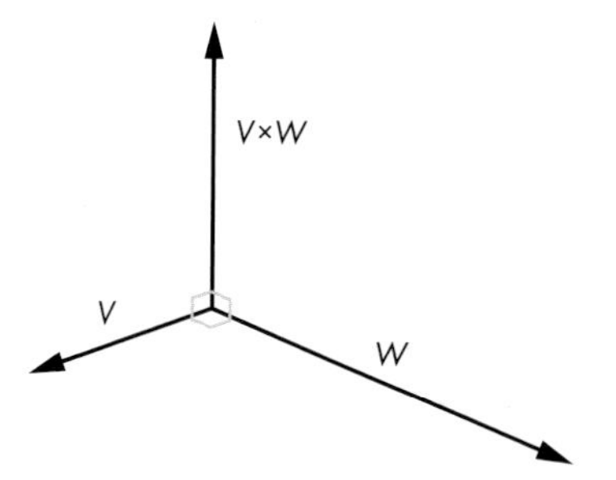

Figure A-7: The cross product of two vectors is a vector perpendicular to both of them.

The computation is a bit more involved than the dot product. If $\vec{R} = \vec{V} \times \vec{W}$, then

$$R_x = V_y \cdot W_z - V_z \cdot W_y$$

$$R_y = V_x \cdot W_z - V_z \cdot W_x$$

$$R_z = V_x \cdot W_y - V_y \cdot W_x$$

The cross product is not commutative. Specifically, $\vec{V} \times \vec{W} = -(\vec{W} \times \vec{V})$.

We use the cross product to compute the *normal vector* of a surface—that is, a unit vector perpendicular to the surface. To do this, we take two vectors on the surface, calculate their cross product, and normalize the result.

Matrices

A *matrix* is a rectangular array of numbers. For the purposes of this book, matrices represent *transformations* that can be applied to points or vectors, and we refer to them with a capital letter, such as M. This is the same way we refer to points, but it will be clear by the context whether we're talking about a matrix or a point.

A matrix is characterized by its size in terms of columns and rows. For example, this is a 4×3 matrix:

$$\begin{pmatrix} 1 & 2 & 3 & 4 \\ -3 & -6 & 9 & 12 \\ 0 & 0 & 1 & 1 \end{pmatrix}$$

Matrix Operations

Let's see what we can do with matrices and vectors.

Adding Matrices

You can add two matrices, as long as they have the same size. The addition is done element by element:

$$\begin{pmatrix} a & b & c \\ d & e & f \\ g & h & i \end{pmatrix} + \begin{pmatrix} j & k & l \\ m & n & o \\ p & q & r \end{pmatrix} = \begin{pmatrix} a+j & b+k & c+l \\ d+m & e+n & f+o \\ g+p & h+q & i+r \end{pmatrix}$$

Multiplying a Matrix by a Number

You can multiply a matrix by a number. You just multiply every element of the matrix by the number:

$$n \cdot \begin{pmatrix} a & b & c \\ d & e & f \\ g & h & i \end{pmatrix} = \begin{pmatrix} n \cdot a & n \cdot b & n \cdot c \\ n \cdot d & n \cdot e & n \cdot f \\ n \cdot g & n \cdot h & n \cdot i \end{pmatrix}$$

Multiplying Matrices

You can multiply two matrices together, as long as their sizes are compatible: the number of columns in the first matrix must be the same as the number of rows in the second matrix. For example, you can multiply a 2×3 matrix by a 3×4 matrix, but not the other way around! Unlike numbers, the order of the multiplication matters, even if you're multiplying together two square matrices that could be multiplied in either order.

The result of multiplying two matrices together is another matrix, with the same number of rows as the left-hand side matrix, and the same number of columns as the right-hand side matrix. Continuing with our example above, the result of multiplying a 2×3 matrix by a 3×4 matrix is a 2×4 matrix.

Let's see how to multiply two matrices, A and B:

$$A = \begin{pmatrix} a & b & c \\ d & e & f \end{pmatrix}$$

$$B = \begin{pmatrix} g & h & i & j \\ k & l & m & n \\ o & p & q & r \end{pmatrix}$$

To make things more clear, let's group the values in A and B into vectors: let's write A as a column of row (horizontal) vectors and B as a row of column (vertical) vectors. For example, the first row of A is the vector (a, b, c) and the second column of B is the vector (h, l, p):

$$A = \begin{pmatrix} (a, b, c) \\ (d, e, f) \end{pmatrix}$$

$$B = \begin{pmatrix} \begin{pmatrix} g \\ k \\ o \end{pmatrix} & \begin{pmatrix} h \\ l \\ p \end{pmatrix} & \begin{pmatrix} i \\ m \\ q \end{pmatrix} & \begin{pmatrix} j \\ n \\ r \end{pmatrix} \end{pmatrix}$$

Let's give names to these vectors:

$$A = \begin{pmatrix} -\vec{A_0}- \\ -\vec{A_1}- \end{pmatrix}$$

$$B = \begin{pmatrix} | & | & | & | \\ \vec{B_0} & \vec{B_1} & \vec{B_2} & \vec{B_3} \\ | & | & | & | \end{pmatrix}$$

We know that A is 2×3, and B is 3×4, so we know the result will be a 2×4 matrix:

$$\begin{pmatrix} -\vec{A_0}- \\ -\vec{A_1}- \end{pmatrix} \cdot \begin{pmatrix} | & | & | & | \\ \vec{B_0} & \vec{B_1} & \vec{B_2} & \vec{B_3} \\ | & | & | & | \end{pmatrix} = \begin{pmatrix} c_{00} & c_{01} & c_{02} & c_{03} \\ c_{10} & c_{11} & c_{12} & c_{13} \end{pmatrix}$$

Now we can use a simple formulation for the elements of the resulting matrix: the value of the element in row r and column c of the result—that is, c_{rc}—is the dot product of the corresponding row vector in **A** and column vector in **B**, that is, $\vec{A_r}$ and $\vec{B_c}$:

$$\begin{pmatrix} -\vec{A_0}- \\ -\vec{A_1}- \end{pmatrix} \cdot \begin{pmatrix} | & | & | & | \\ \vec{B_0} & \vec{B_1} & \vec{B_2} & \vec{B_3} \\ | & | & | & | \end{pmatrix} = \begin{pmatrix} \langle \vec{A_0}, \vec{B_0} \rangle & \langle \vec{A_0}, \vec{B_1} \rangle & \langle \vec{A_0}, \vec{B_2} \rangle & \langle \vec{A_0}, \vec{B_3} \rangle \\ \langle \vec{A_1}, \vec{B_0} \rangle & \langle \vec{A_1}, \vec{B_1} \rangle & \langle \vec{A_1}, \vec{B_2} \rangle & \langle \vec{A_1}, \vec{B_3} \rangle \end{pmatrix}$$

For example $c_{01} = \langle \vec{A_0}, \vec{B_1} \rangle$, which expands to $a \cdot h + b \cdot l + c \cdot p$.

Multiplying a Matrix and a Vector

You can think of an n-dimensional vector as either an $n \times 1$ vertical matrix or as a $1 \times n$ horizontal matrix, and multiply the same way you would multiply any two compatible matrices. For example, here's how to multiply a 2×3 matrix and a 3D vector:

$$\begin{pmatrix} a & b & c \\ d & e & f \end{pmatrix} \cdot \begin{pmatrix} x \\ y \\ z \end{pmatrix} = \begin{pmatrix} a \cdot x + b \cdot y + c \cdot z \\ d \cdot x + e \cdot y + f \cdot z \end{pmatrix}$$

Since the result of multiplying a matrix and a vector (or a vector and a matrix) is also a vector and, in our case, matrices represent transformations, we can say that the matrix *transforms* the vector.

INDEX

G

H

I

K

L

M

N

O

P

Q

R

S

T

U

V

W

X

Y

Z

The fonts used in *Computer Graphics from Scratch* are New Baskerville, Futura, The Sans Mono Condensed, and Dogma. The book was typeset with LaTeX 2ε package `nostarch` by Boris Veytsman *(2008/06/06 v1.3 Typesetting books for No Starch Press).*

The book was printed and bound by Versa Printing, Inc. in East Peoria, Illinois. The paper is 70# Matte, which is certified by the Forest Stewardship Council (FSC).

The book uses a layflat binding, in which the pages are bound together with a cold-set, flexible glue and the first and last pages of the resulting book block are attached to the cover. The cover is not actually glued to the book's spine, and when open, the book lies flat and the spine doesn't crack.